RIP CURRENTS

Beach Safety, Physical Oceanography,
and Wave Modeling

RIP CURRENTS
Beach Safety, Physical Oceanography, and Wave Modeling

EDITED BY
Stephen Leatherman
John Fletemeyer

CRC Press
Taylor & Francis Group
Boca Raton London New York

CRC Press is an imprint of the
Taylor & Francis Group, an **informa** business

The cover photo of Tamarama Beach in Sydney is courtesy of Robert W. Brander of the University of New South Wales, Australia.

CRC Press
Taylor & Francis Group
6000 Broken Sound Parkway NW, Suite 300
Boca Raton, FL 33487-2742

First issued in paperback 2017

ISBN 13: 978-1-4398-3896-9 (hbk)
ISBN 13: 978-1-138-07527-6 (pbk)

Visit the Taylor & Francis Web site at
http://www.taylorandfrancis.com

and the CRC Press Web site at
http://www.crcpress.com

Contents

Preface

This first ever book about rip currents emanated from the First International Rip Current Symposium held at Florida International University in Miami on February 17–19, 2010. More than 100 coastal scientists, engineers, forecast meteorologists, lifeguard chiefs, and other practitioners from ten countries participated in this three-day conference organized and chaired by Dr. Stephen P. Leatherman and Dr. John Fletemeyer.

The overall goal of the First International Rip Current Symposium was to bring together researchers from around the world to identify advancements in rip research that will lead to a better understanding of the dynamics, mechanisms, and predictability of these dangerous currents. In addition, this symposium focused on communicating the rip current threat to the public through various outreach programs and campaigns and evaluating their overall effectiveness. Top academic researchers and National Weather Service forecasters made presentations; a number of these papers constitute the chapters in this book along with other selected papers. Presenters and panelists also included beach managers and lifeguard chiefs because the ultimate objective of this symposium was to reduce the number of rescues, near-drowning incidents, and drownings caused by rip currents.

The 16 chapters in this book span the spectrum of rip current research and outreach initiatives. Chapters on rip studies concern all four U.S. coasts (Atlantic, Gulf, Pacific, and Great Lakes) as well as Brazil, the United Kingdom, Japan, and Australia. Scientific techniques utilized to study rip currents included field investigation and numerical modeling. The field research involved the use of water-based sensors, video technology, and remote sensing.

Clearly the science of rip currents has advanced significantly in the past few years as demonstrated by the chapters in this book, and many more studies are currently underway. The challenge of alerting the public to these dangerous currents is problematic because rip currents come in many sizes, shapes, and strengths. There is no single way to identify rip currents because the water can appear light colored due to entrained sediment and bubbles or dark because of the presence of a channel through which the rips flow when the surf is up.

This book will serve as a primer for rip current research and outreach activities for many years to come and will help chart the agenda for the Second International Rip Current Symposium planned to be held in October 2012 in Sydney, Australia.

Stephen P. Leatherman
Miami, Florida

Acknowledgments

The Andrew W. Mellon Foundation and the National Weather Service are gratefully acknowledged for supporting rip current studies by Stephen Leatherman and making possible the publication of this first book on rip currents.

The Andrew W. Mellon Foundation has supported this coastal processes research program for nearly two decades. This sustained funding allowed Dr. Leatherman, his colleagues and graduate students to pursue a wide range of studies that first focused on coastal erosion. Erosion, both short-term and episodic in nature due to storm impact and long-term changes in response to sediment deficiencies and sea level rise, has been studied utilizing historical shoreline maps and LIght Distance And Ranging (LIDAR).

Rip currents are responsible for significant cross-shore sediment transport. These powerful currents are also the principal threats to bathers and swimmers at surf beaches. In fact, rip currents kill more people on average each year at U.S. beaches than hurricanes, tornadoes, or lightning according to the National Weather Service. The Andrew W. Mellon Foundation funding has facilitated field research and laboratory studies of rip currents. Fluorescent dye tracer studies were conducted to determine the conditions for rip formation and strength along the southeast Florida coast. From these field experiments, Leatherman developed a more convenient means of placing the tracer into the surf zone by formulating the floating dye ball water tracer. The Miami Beach lifeguards have been testing the dye balls and find that they are very useful in detecting rip currents for lifeguard training and for alerting the public of these dangerous currents through demonstrations.

The National Weather Service grant on coastal vulnerability enabled Leatherman and associates to conduct erosion analyses for various Florida beaches and also undertake an outreach program regarding rip currents. In addition to sponsoring the First International Rip Current Symposium (with additional support from Florida Sea Grant), the National Weather Service grant enabled the production of a video titled "Beach Rips: Killer Currents." This video shows an actual rip current in action made visible by a fluorescent green tracer dye; it can be found on YouTube and www.DrBeach.org.

Editors

Dr. Stephen P. Leatherman is a director of the Laboratory for Coastal Research and a professor in the Department of Earth and Environment at Florida International University in Miami. He has published hundreds of journal articles and authored and/ or edited sixteen books on coastal science. In 2003, he authored a Yale University Press book on beach safety with emphasis on rip currents. Dr. Leatherman has given expert testimony to U.S. Congressional committees eleven times and has appeared five times in yearbooks of *Who's Who in America* and *Who's Who in the World*. He served as the on-screen host and co-producer of the "Vanishing Lands" documentary that won three international awards. His preferred name bestowed by the media is "Dr. Beach"—the name under which he releases annually the highly acclaimed list of America's Top Ten Beaches based on fifty criteria.

Dr. John R. Fletemeyer was the chief of the Palm Beach lifeguards for sixteen years and has been involved in professional aquatics for more than three decades. He is the recipient of many national awards including the Thirty-Year National Association of Underwater Instructors Service Award and the Top One Hundred Aquatic Professionals in America Award. Dr. Fletemeyer has served in a number of leadership positions including chairman of the National Aquatic Coalition, national education chairman and president of the United States Lifesaving Association Southeast Region, and chairman of the board of the International Swimming Hall of Fame. He is the author of many aquatic safety articles and was co-editor of *Drowning: New Perspectives in Intervention and Prevention* published by CRC Press.

Drs. Leatherman and Fletemeyer were assisted by **Dr. Ivonne Schmid** of the International Swimming Hall of Fame.

Contributors

Miguel da G. Albuquerque
Instituto Federal de Educação
Ciência e Tecnologia do Rio Grande
   do Sul
Porto Alegre, Brazil

M. J. Austin
School of Marine Science and
   Engineering
University of Plymouth
Plymouth, United Kingdom

Henry Bokuniewicz
School of Marine and Atmospheric
   Sciences
Stony Brook University
Stony Brook, New York

Robert W. Brander
School of Biological, Earth, and
   Environmental Sciences
University of New South Wales
Sydney, Australia

Nicole Caldwell
Department of Environmental Studies
University of West Florida
Pensacola, Florida

Lauro J. Calliari
Universidade Federal do Rio Grande
   do Sul
Porto Alegre, Brazil

Gene Clark
Sea Grant Institute
Lake Superior Field Office
University of Wisconsin
Superior, Wisconsin

Robert A. Dalrymple
Civil Engineering Department
Johns Hopkins University
Baltimore, Maryland

Robert G. Dean
Department of Civil and Coastal
   Engineering
University of Florida
Gainesville, Florida

Gregory Dusek
Department of Marine Sciences
University of North Carolina
Chapel Hill, North Carolina

David Elder
Ocean Rescue Unit
Town of Kill Devil Hills
Kill Devil Hills, North Carolina

John R. Fletemeyer
Aquatic Research, Conservation and
   Safety (ARCS)
Fort Lauderdale, Florida

Paul Gayes
Coastal Carolina University
Center for Marine and Wetland Studies
Conway, South Carolina

David Guenther
National Weather Service
Marquette, Michigan

Jeffrey Hanson
U.S. Army Corps of Engineers
Field Research Facility
Duck, North Carolina

Brian K. Haus
Rosenstiel School of Marine and
    Atmospheric Science
University of Miami
Miami, Florida

Chris Houser
Department of Geography
Texas A&M University
College Station, Texas

Ronald E. Kinnunen
Sea Grant Extension
Michigan State University
Marquette, Michigan

Antonio Henrique da F. Klein
Universidade do Vale do Itajaí.
Itajai, Brazil

Nicholas C. Kraus
U.S. Army Engineer Research and
    Development Center
Coastal and Hydraulics Laboratory
Vicksburg, Mississippi

Nirnimesh Kumar
Department of Earth and Ocean
    Sciences
University of South Carolina
Columbia, South Carolina

Stephen P. Leatherman
Laboratory for Coastal Research
Florida International University
Miami, Florida

James B. Lushine
Formerly with National Weather
    Service
Pembroke Pines, Florida

Jamie H. MacMahan
Oceanography Department
Naval Postgraduate School
Monterey, California

Gerd Masselink
School of Marine Science and
    Engineering
University of Plymouth
Plymouth, United Kingdom

Guy Meadows
Department of Naval Architecture and
    Marine Engineering
University of Michigan
Ann Arbor, Michigan

Lorelle Meadows
College of Engineering
University of Michigan
Ann Arbor, Michigan

Klaus Meyer-Arendt
Department of Environmental Studies
University of West Florida
Pensacola, Florida

Onir Mocellin
Corpo de Bombeiros do Estado de
    Santa Catarina
Itajai, Brazil

Varjola Nelko
Civil Engineering Department
Johns Hopkins University
Baltimore, Maryland

Charles H. Paxton
National Weather Service
Ruskin, Florida

Heidi Purcell
Department of Naval Architecture and
    Marine Engineering
University of Michigan
Ann Arbor, Michigan

Paul Russell
School of Marine Science and
    Engineering
University of Plymouth
Plymouth, United Kingdom

Timothy M. Scott
School of Marine Science and
    Engineering
University of Plymouth
Plymouth, United Kingdom

Harvey Seim
Department of Marine Sciences
University of North Carolina
Chapel Hill, North Carolina

Michael P. Slattery
School of Marine and Atmospheric
    Sciences
Stony Brook University
Stony Brook, New York

R. J. Thieke
Department of Civil and Coastal
    Engineering
University of Florida
Gainesville, Florida

George Voulgaris
Department of Earth and Ocean
    Sciences
University of South Carolina
Columbia, South Carolina

John C. Warner
Woods Hole Oceanographic Institution
Coastal and Marine Geology Program
Woods Hole, Massachusetts

S. Wills
Royal National Lifeboat Institution
Poole, United Kingdom

A. Wooler
Royal National Lifeboat Institution
Poole, United Kingdom

1 Future Challenges for Rip Current Research and Outreach

Robert W. Brander and Jamie H. MacMahan

CONTENTS

THE RIP CURRENT PROBLEM

Since recreational beach swimming became popular in the early 1900s, many unfortunate beachgoers have drowned in rip currents—strong, narrow, and concentrated flows of water that are common on many ocean, inland sea, and lacustrine beaches characterized by breaking waves. Rip currents can quickly carry unsuspecting bathers in an offshore direction against their will, where a combination of lack of knowledge, panic, and exhaustion too often leads to serious consequences. Not surprisingly, rip currents are the leading causes of beach rescues and drownings.

Rip currents clearly represent a serious public health issue with major personal, societal, and economic costs associated with drowning deaths, near-miss drownings, injuries, and trauma (Sherker et al., 2008). The United States Lifesaving Association (USLA) estimates that rip currents account for 80% of surf rescues and the annual number of fatalities due to rip currents exceeds 100. However, other estimates vary from as low as 35 (Gensini and Ashley, 2009) up to 150 (Lushine, 1991). From 1960 to 2000, the total cost of drowning deaths at American beaches was estimated at U.S. $4.2 billion (Branche and Stewart, 2001). In Australia, 89% of the more than 25,000 annual surf rescues are caused by rip currents (Short and Hogan, 1994) with an estimated 40 to 50 drownings per year (Sherker et al., 2008; SLSA, 2009). Other countries and regions with rip current problems include New Zealand, United Kingdom, Europe, Israel, the Middle East, South Africa, Asia, and Central and South America (Klein et al., 2003; Carey and Rogers, 2005; Hartmann, 2006; Bech, 2007; Short, 2007; McCool et al., 2008; Scott et al., 2009). Rip currents clearly constitute a global hazard.

Unfortunately, due to a lack of national reporting systems and also to the logistical difficulties involved in obtaining accurate and reliable incident reports, we simply do not know the actual number of people who drown in rip currents. This is particularly true for many developing countries where anecdotal reports suggest that the incidence of rip current drownings on recreational beaches may be extremely high. Nevertheless, even using conservative estimates, it is probable that the annual number of drownings in rip currents worldwide exceeds 500. This figure would be much higher without the beach lifeguard services and public rip current safety education that have both increased over time. However, lifeguards are not present on all beaches at all times, so many beachgoers simply lack this safety net or choose to ignore it.

It is difficult to assess long-term trends in the incidence of rip current drownings, primarily due to the lack of accurate data, but also due to the impacts of weather and economic conditions on beach visitation numbers, and the effects of surf conditions on the numbers of bathers and the numbers and intensities of rip currents. Both types of impacts can be extremely variable over days, seasons, and years. The human factor involving when, where, and why beachgoers go into the ocean adds further complexity to the problem. For these reasons, rip current drownings fluctuate significantly from year to year and over longer cycles, making it difficult to gauge the effectiveness of beach lifeguards, public rip current education, and scientific advances focused on mitigating rip hazards. Nevertheless, despite our collective efforts, there is little evidence to suggest that the number of rip current drownings in recent decades is decreasing, even in nations such as the United States (Gensini and Ashley, 2009) and Australia (SLSA, 2009) that have pioneered lifesaving measures and public rip education. Clearly something is not working properly.

Our basic understanding of rip current behavior has also improved from increased laboratory, numerical modeling, and field studies. Rip currents have long served as the focus of scientific attention as they are integral components of near-shore cell circulation, play a crucial role in the morphodynamic evolution of beaches and surf zones, and contribute significantly to beach erosion, sediment transport, and nutrient and pollutant dispersal. Consequently, since the early descriptive work on rip currents in the 1920s, our accumulated scientific knowledge and associated understanding of rip currents has steadily increased. The sizes, intensities, and characteristics of rip currents vary from beach to beach, over tidal cycles, and over different wave conditions (MacMahan et al., 2006). This complexity creates difficulty in describing a

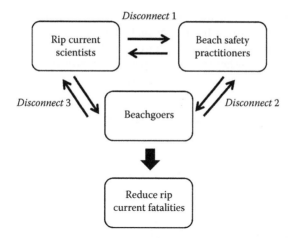

FIGURE 1.1 Transfer and translation of knowledge among rip current scientists, beach safety practitioners, and the beach-going public are hindered by several information disconnects.

universal rip current definition and the result is often a single idealized diagram that does not properly address the true behavior of a rip current for a particular setting.

Coastal scientists and beach safety practitioners are well aware of the hazards that rip currents represent to beachgoers. Unfortunately, most beachgoers are unaware of the various surf zone processes that can be harmful to them. This fundamental lack of awareness arises from several disconnects that have hindered the ability to adequately translate basic scientific knowledge of rip currents to the average beachgoer (Figure 1.1).

Traditionally the role of rip current science (and scientists) has been the attainment of knowledge and information relating to the behaviors of rip current systems. While some of this information may reach beachgoers directly, it is usually up to beach safety practitioners to disseminate it and determine the best methods to apply this information to educate the beach-going public. At the same time, they help advise rip current scientists to attain information that is useful to their education programs.

The first disconnect exists because little scientific information is relevant or readily accessible outside the scientific community and, until relatively recently, few collaborations between rip current scientists and beach safety practitioners allowed concerted two-way communications and development of appropriate rip current science. A second disconnect relates to lack of education. Despite ongoing rip current outreach efforts in many areas, people continue to be rescued from or drown in rips at alarming rates. Several studies (Ballantyne et al., 2005; Sherker et al., 2010) have shown that beachgoers remain largely ignorant of rip currents. The third disconnect is that science has largely ignored what beachgoers can tell us about rip currents. The challenge for the rip current community at large is to improve the flow of information and subject it to quality control in an effort to reduce rip current fatalities (Figure 1.1).

In an attempt to foster rip current safety, this chapter provides an overview of (1) scientific rip current knowledge that is directly relevant to beach safety practitioners; and (2) community outreach strategies promoting rip current education and awareness. Existing limitations and challenges facing rip current science and outreach programs are identified and we provide suggestions that may hopefully lead to an improved translation of basic rip current knowledge to beachgoers and a reduction in rip current incidents in the future.

RIP CURRENT SCIENCE AND BEACH SAFETY

TRADITIONAL RIP CURRENT PARADIGM

The "rip current" term was coined by Shepard (1936) to distinguish these offshore flows from "undertows" and "rip tides" that were popular, but conceptually misleading terms commonly used in the literature and public vernacular at the time (Davis, 1925). Anecdotal evidence suggests that many beachgoers perceive rip currents as undertows that will pull them under the water. In reality, rip current flow does not pull people or water below the surface. Rip currents actually flow fastest near the surface (Haas and Svendsen, 2002). "Rip tide" is a technically incorrect term because rip currents are not tides, and tidal rips are different currents associated with tidal inlets during an ebbing tide. Unfortunately, both terms are still commonly and incorrectly used by the media and general public to describe rip currents. The term "undertow" also remains in use by coastal scientists, but describes a laterally homogeneous current that flows offshore near seabeds at velocities much lower than those of rip currents (Garcia-Faria et al., 2000; Aagaard and Vinther, 2008). Undertow commonly occurs on beaches with relatively minimal alongshore variation in sandbar relief. Rip currents tend to occur on beaches that have pronounced alongshore variations in sandbar relief.

The first serious scientific attempts to describe rip currents came from the Scripps Institution of Oceanography at La Jolla, California (Shepard et al., 1941; Shepard and Inman, 1950, 1951; Inman and Quinn, 1952), and the foundations of our conventional understanding of rip current systems effectively originated from these studies. The fundamental paradigm is that rip currents are key components of an idealized near-shore circulation cell (Figure 1.2) involving a continuous interchange of water between the surf zone and areas offshore (Inman and Brush, 1973). They consist of alongshore feeder currents fully contained within the surf zone, but restricted close to shore, that carry water toward a narrow and fast-flowing shore-normal rip neck that extends through the surf zone. Rip current flow then extends well beyond the surf zone where it decelerates as an expanding rip head. This water is then able to return shoreward through the action of waves completing the cell.

The description of rip currents is often overly simplified by a mass balance description. Waves transport water shoreward and the water piles up at the shoreline, requiring an offshore balance that can be achieved by rip currents. This is partially correct. Rip currents are driven by gradients in wave momentum, referred to as radiation stresses (Longuet-Higgins and Stewart, 1964). These gradients occur in both the cross-shore and alongshore directions. The cross-shore momentum gradients are the result of wave shoaling and breaking (Bowen, 1969). Most exposed beaches have alongshore variations in sandbar relief and surf zone morphology (Lippmann and Holman, 1989) that induce alongshore variations in wave breaking (radiation stresses).

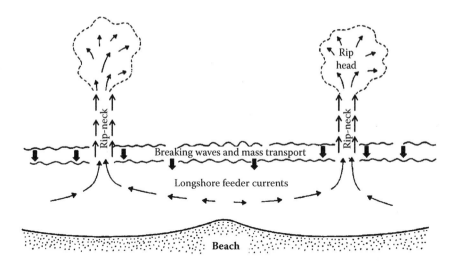

FIGURE 1.2 Traditional paradigm of rip current circulation typically shows flow extending well beyond the surf zone. (*Source:* Modified from Komar, 1998. *Beach Processes and Sedimentation*, 2nd ed. Prentice Hall, New York.)

Larger waves break in deeper water and smaller waves break in shallower water. Assuming similar wave conditions approach a beach with alongshore variations in sandbar morphology, alongshore variations in wave breaking will occur and drive rip currents (Bowen, 1969; Dalrymple, 1978). In simple terms, water flow moves from regions of intense wave breaking to regions with less wave breaking that are often deeper. However, the flows are not simply alongshore and then offshore (Figure 1.2). In general, rip currents are clockwise and counter-clockwise rotating eddies that vary in velocity magnitude, shape, and orientation (Reniers et al., 2009) rather than simple offshore-directed jets.

The idealized view of a rip system (Figure 1.2) as defined by Shepard et al. (1941) appears in various forms in almost every popular coastal science textbook (Komar, 1998; Dean and Dalrymple, 2002; Masselink and Hughes, 2003; Davis and Fitzgerald, 2004) and, not surprisingly, has been incorporated by beach safety practitioners in most rip current education programs. Of note, the traditional rip current paradigm is based on early "visual" observations made on beaches of convenience where rip currents were strongly influenced by piers or submarine canyons (Shepard and Inman, 1950) that forced rip flows considerable distances seaward of the surf zone. In reality, these rip currents can be considered anomalous relative to the much more common rips that develop along open coast beaches, have surf zone morphodynamic associations, are modified by directional waves, and are not controlled by structures or offshore topography. Furthermore, results from recent field experiments on open coast rip currents by MacMahan et al. (2010) are now challenging the traditional paradigm shown in Figure 1.2, suggesting that our conventional understanding of rip cell circulation may represent the exception rather than the rule.

The new paradigm shift has largely been associated with the development, modification, and use of Global Positioning System (GPS) devices attached to drifters

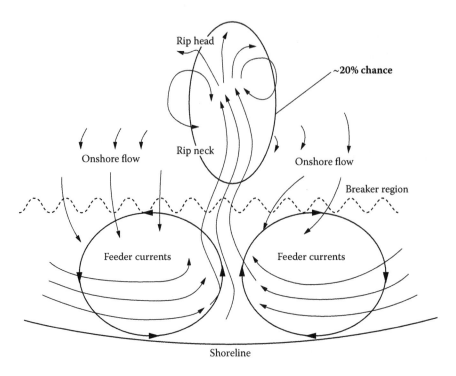

FIGURE 1.3 Modified view of rip current circulation based on recent field measurements by MacMahan et al. (2010) indicating that most rip currents exhibit circulatory behavior fully contained within the surf zone with only 20% of water exiting the surf zone.

and deployed en masse in rip current systems (MacMahan et al., 2009). This relatively simple Lagrangian methodology has dramatically increased our spatial coverage of rip current flow measurements to provide more holistic observations and analysis of system flow behavior. Drifter measurements on beaches in California, England, and France (MacMahan et al., 2010) that have similar surf zone rip channel morphologies (but different wave and tidal regimes) showed that cellular rip current circulation is maintained within the surf zone approximately 80% of the time with only 20% of drifters and water actually exiting the surf zone (Figure 1.3). Importantly, these results show that the eddy-like nature of rip current flow in the surf zone is often masked by breaking waves. When a rip current exits from the surf zone, it diffuses into quiescent water, making the flow much easier to recognize. Other studies (Brander and Short, 2001; Austin et al., 2010) revealed that lateral flow of water across shallow sand bars may in fact contribute more water to the rip neck than the idealized alongshore feeder currents shown in Figure 1.2.

These findings have major implications for both beach safety practitioners and beachgoers. They suggest an 80% chance that bathers caught in rip currents who simply stay afloat and tread water will be recirculated back to the relative safety of shallow sandbars within minutes without leaving the surf zone. The findings also indicate some potential flaws with the common advice to escape a rip by "swimming parallel to the beach." Depending on the direction in which a swimmer chooses to swim, there is a chance he will be swimming against a strong alongshore drift of water. Understandably, these results have been met with skepticism from beach safety practitioners because the results significantly challenge traditional wisdom. It has also been suggested that the new findings are applicable only to the types of rips on the beaches studied and do not accurately reflect rip current behaviors for all environments. For this reason, it is useful to review what is known about the different types of rip currents.

TYPES AND IDENTIFICATION OF RIPS

It is often assumed that it is difficult, if not impossible, to educate the general public on how to identify a rip current (Carey and Rogers, 2005; Fletemeyer and Leatherman, 2010) because different types of rip currents vary in intensity, location, and appearance, depending on morphological and hydrodynamic conditions. However, as noted by Short (2007), the most common types on beaches are open coast "fixed rips" occupying distinct channels that persist in location for days, weeks, and even months in association with transverse bar and rip morphologies and modal wave conditions (Figure 1.4).

As fixed rips occur under decreasing or extended periods of low to moderate wave energy conditions, they usually appear as dark, clear gaps between areas of whitewater due to less frequent wave breaking over deeper rip channels (Figure 1.5). While this characteristic provides a fairly simple visual identification clue, it is often deceptive and dangerous for inexperienced beach users who actively choose to swim in the "calmer" water away from breaking waves (Sherker et al., 2010). Anecdotally, fixed rip currents are also considered by lifeguards to be the most dangerous types of rip currents as they exist during ideal weather that promotes large beach crowds, and less energetic surf conditions that are more attractive for swimming. Fixed rips also tend to recur along beaches, with quasi-regular alongshore spacings ranging from 50 m to more than 500 m as wave energy increases (Short and Brander, 1999; Thornton et al., 2007).

Another type of fixed rip current occurs adjacent to artificial structures such as groins, jetties, and piers, or natural features like headlands and reefs. In both cases, the rip current flow is topographically constrained in location, creating an almost permanent channel that becomes active during periods of wave breaking. Short (2007) suggests that these "topographic rips" have stronger, more confined flows than open beach fixed rips and therefore carry flows (and swimmers) greater distances seaward, although Pattiaratchi et al. (2009) found persistent rip recirculation in the lee of a groyne.

Most topographic rips are also commonly identified as darker, calmer gaps between regions of breaking waves (Figure 1.6) and therefore have the same inherent risk of appearing safer to inexperienced swimmers, but with additional risks posed by collisions with solid features. Fixed topographic rips exist and are maintained under both modal and high energy wave conditions (Short, 2007). During extreme wave energy events, topographic rips may flow hundreds of meters offshore and are referred to as "megarips." The seaward extent of a megarip can be identified by plumes of sediment in the expanding and decelerating rip head.

(a)

(b)

FIGURE 1.4 Examples of open-coast fixed beach rip currents: (a) Shelly Beach, New South Wales, Australia (Courtesy of Wyong Shire Council); (b) Pensacola Beach, Florida (Courtesy of Bob West); (c) Truc Vert, France (Courtesy of Nadia Senechal); (d) Marina Beach, California (Courtesy of Robert W. Brander).

Open coast beach rips may also be temporally and spatially transient in nature, lasting only several minutes at a particular location or migrating slowly along the shore. These "flash rips" are hydrodynamically rather than topographically forced. They are associated with incident wave conditions characterized by different frequencies and directions, dissipative surf zone conditions, and wave groups, all of

(c)

FIGURE 1.4 (continued).

which contribute to variable wave set-up gradients (Tang and Dalrymple, 1989; Fowler and Dalrymple, 1990; Reniers et al., 2009; MacMahan et al., 2010). They typically appear as narrow bands of choppy, churning, water through the surf zone with streaky, turbulent and sediment laden rip head plumes just seaward of the surf zone (Figure 1.7).

RIP FLOW CHARACTERISTICS

Field measurements of rip currents are logistically difficult to obtain, but existing measurements from a range of rip environments from beaches around the world show consistent patterns of flow behavior that are of direct relevance to beach and rip safety. Measurements of open coast beach fixed rips under low to moderate wave energy conditions indicate that rip flow is tidally modulated, flowing faster around low tide, with mean velocities typically on the order of 0.3 to 0.8 ms^{-1}, decreasing in strength toward high tide (Sonu, 1972; Bowman et al., 1988; Aagaard et al., 1997;

(d)

FIGURE 1.4   (continued).

FIGURE 1.5   (*See color insert.*) Many open-coast fixed beach rip currents appear as "dark gaps" of "calm" water between areas of breaking waves. To inexperienced beachgoers, they often look like the safest places to swim. (Courtesy of Rob Brander.)

FIGURE 1.6 The famous "Backpacker Express" along the southern headland at Bondi Beach in Sydney is a good example of a structurally fixed rip. (Courtesy of Robert W. Brander.)

FIGURE 1.7 Flash rip on Rehoboth Beach in Delaware is evident by the turbulent rip head plume extending seaward from the surf zone. (Courtesy of Wendy Carey and Delaware Sea Grant.)

Brander and Short, 2000; MacMahan et al., 2005). Flow velocities of transient rips have been shown to be low, on the order of 0.2 to 0.3 ms^{-1} (MacMahan et al., 2010). However, while this information has important implications to rip current safety, it is largely unknown and meaningless to beachgoers.

It is often reported in educational materials that rip currents flow faster than most swimmers and indeed flow faster than Olympic swimmers. However, these statements are misleading and not entirely true. Competitive swim times for the 50-m freestyle equate to swimming speeds of 2.3 ms^{-1}. Average swimmers swim easily at speeds of 1 ms^{-1}. In both cases, the swimmers could hypothetically swim against low to moderate energy open-coast beach rips back to shore or to adjacent shallow sand bars, over short distances before being carried further offshore. Only during very brief rip pulsations, when sudden and brief accelerations of 2ms^{-1} or more may occur (Brander and Short, 2001; MacMahan et al., 2004) are average swimmers carried seaward of the surf zone. Not surprisingly, rip current pulses are considered major causes of mass rescues of swimmers (Short, 2007). Megarips and rip currents in high energy environments also pose great risks, as mean flows may exceed 2 ms^{-1} over larger length scales (Brander and Short, 2000; Short, 2007). The implication here is that non-swimmers and poor swimmers face the greatest risk. However, this argument does not consider swimmer behavior and response in an unfamiliar situation. Even strong swimmers can panic when caught in rip currents.

RIP PREDICTION AND FORECASTING

Rip current forecasts come in different forms. Correlative predictions attempt to relate the occurrence of rip current rescues with weighted environmental meteorological and oceanographic conditions such as wind speed, direction, tidal level, wave height, and period (Lushine, 1991; Lascody, 1998; Engle et al., 2002). These are formulated to provide rip current hazard scales and predictive indices, but are often not normalized with beach visitation numbers. For example, a high rip current danger level may occur when no one is on a beach—resulting in zero rescues—while a low danger level during a crowded period may be accompanied by a high number of rescues. The human factor also plays a significant role. Beachgoers may or may not enter the ocean for a number of reasons including air and water temperature, cloud cover, wind, and jellyfish that have no relation to rip current activity. This makes it difficult to assess the predictability of rip current hazard. Nevertheless, owing to the simplicity of the predictions, correlative predictions represent the most common approach (Carey and Rogers, 2005).

However, rip current activity varies from beach to beach and should be evaluated based on location. Lifeguard records in the United States show that rip current rescues increase with increasing wave height, stop when waves get too big, and increase during lower tides; this is consistent with in situ field observations. In Daytona Beach, Florida, there is a 50% chance of rip current rescue per day and the number of rip current rescues increases with increasing beach visitation (Engle et al., 2002). Significant numbers of rip current rescues (known as outbreaks) occur on certain days and drownings tend to occur on these days. Engle et al. (2002) found directional wave spreading, the variation in the angle of wave approach, to be the sole metric for determining when rip current outbreaks occur. Rip currents are enhanced when waves come primarily from a shore-normal direction. Unfortunately, most beaches in the U.S. do not have directional wave sensors that would dramatically improve rip current predictions. More investigation is required to determine when rescue outbreaks occur so that the lifesaving community can be prepared.

Predicting the formation, spacing, and location of rip currents is a significant scientific challenge that has largely been the focus of numerical and laboratory modeling studies (Chen et al., 1999; Haas and Svendsen, 2002; Reniers et al., 2009). Improved numerical models better replicate the complicated circulation patterns of rip currents, but require bathymetry and offshore directional wave data that may not be available. By modifying the surf zone morphology and input wave conditions, numerical models can expand our present understanding of physical processes that are too difficult or costly to analyze in the field. The transition of applying numerical models to beach communities has been limited by lack of funds and trained personnel to perform numerical simulations. However, numerical models of wave prediction are used for many beaches and a few sites provide some surf zone current model predictions.

The application of relatively inexpensive remote video camera imagery and associated image processing techniques in the late 1980s (Lippmann and Holman, 1989) showed that most beaches support rip currents that can persist for many months. Furthermore, video imagery has highlighted the complexities associated with sandbar evolution. These findings underscore the fundamental issue with rip currents: they cannot be described by a simple idealized diagram. Analysis of long-term (years) video imagery of rip current morphology on beaches with differing wave climates and tidal regimes has shown that (1) rip spacing is more irregular than regular, (2) rip channels can migrate alongshore, and (3) the most common morphological expression of rip current channels is associated with a transverse bar-rip beach state (Ranasinghe et al., 2000; 2004; Holman et al., 2006; Turner et al., 2007).

Significant uncertainty remains regarding the location and persistence of rip channels after major storm events as well as the role of antecedent morphology and wave energy. Monitoring the occurrence and behavior of flash rips using video presents a further challenge. MacMahan et al. (2008) found that subtle alongshore variations in bathymetry can generate rip currents that may be assumed to be flash rips. The ability to observe rip current locations in real time through the use of publicly available online video imagery clearly has potential safety applications both for lifeguards and beachgoers with access to the Internet. However, the logistics and costs of establishing cameras providing adequate coverage along so many popular recreational beaches are enormous. Experienced lifeguards are equally adept at identifying the locations of fixed rip currents on beaches through direct observation.

Since the mid-1980s, the National Weather Service (NWS) of the U.S. National Oceanographic and Atmospheric Administration (NOAA) has utilized the correlative rip current predictive indices of Lushine (1991) and Lascody (1998) to issue surf zone forecasts and rip current outlooks directly to the public as part of daily weather forecasts on television and from weather stations (Carey and Rogers, 2005). NWS offices along the eastern seaboard, Gulf Coast, and in California, Hawaii, and Guam issue rip current outlooks based on a low-to-moderate-to-high risk scale. While uncertainty remains as to the accuracy and efficacy of the predictions, the incorporation of rip current warnings in a public forum is a positive step for introducing the "rip current" terminology to the public. However, the impact and effectiveness of issuing such warnings on the frequency of surf rescues and overall awareness of beachgoers to rip current hazard in the U.S. has yet to be assessed.

RIP CURRENT OUTREACH

SIGNAGE, FLAGS, AND LIFEGUARDS

Traditionally, the most primary form of rip current hazard mitigation involves warning signs and beach lifeguard services. Often erected for legislative and liability prevention purposes, rip current signage can vary greatly by location and among local, state, and national government agencies (Figure 1.8). Signs are usually posted near public access points at both patrolled and unpatrolled beaches. In general, rip current signs consists of text-based warnings that may or may not use the term "rip current" and lack educational value (Figure 1.8a) or more informative signs incorporating text and a diagram showing the basic anatomy of a rip current system and how to react if caught in one (Figure 1.8b). More elaborate diagrammatic signs include basic explanations of rip currents, tips on how to identify them, and detailed text instructions on how to respond if caught in them (Figure 1.8c). Unfortunately, no formal and rigorous assessment of the effectiveness of rip current signage has been published. We simply do not know how many people read, understand, or follow the basic information presented on rip current signs or the impacts of the signs on numbers of rip rescues and drownings.

(a)

FIGURE 1.8   Variabilities in rip current signage are evident from these examples at beaches. (a) Australia. (Courtesy of Matthew Celia) (b) Hawaii (www.panaramio.com). (c) North Carolina. (Courtesy of Spencer Rogers)

(b)

FIGURE 1.8 (continued).

In most countries, the presence of beach lifeguards is obvious through lifeguard towers, vehicles, and beach flags, but these identifiers vary around the world. The U.S. uses a colored flag and sign system that indicates the relative safety level of the surf. Australia, New Zealand, the United Kingdom, and other countries use a pair of international safety standard red and yellow flags (George, 2007) on beaches and constantly emphasize the instruction to "swim between the red and yellow flags." While such systems may be effective locally (Klein et al., 2003), not all beachgoers chose to swim in the proximity of lifeguards, and the global variations in beach flags globally may create confusion for international travelers.

The value of lifeguards in reducing the incidence of rip current drownings is immense. In the U.S., the chance of death by drowning on beaches patrolled by USLA lifeguards is 1 in 18 million (Branche and Stewart, 2001). In Australia, most of the hundreds of rip current drownings over the last 10 years occurred on unpatrolled beaches or outside of supervised areas and lifeguard patrol times (SLSA, 2009) on patrolled beaches. However, it is logistically and economically impossible to provide lifeguarding services on all beaches all the time. Furthermore, patrolled areas on beaches represent only a very small area and most beaches in temperate climates are

(c)

FIGURE 1.8   (continued).

patrolled only seasonally during warmer swimming months. While more lifeguards are clearly the best solution, they simply cannot be everywhere all the time.

EDUCATION AND AWARENESS STRATEGIES

Largely in response to drownings and the fact that lifeguard presence is not guaranteed, most regions with popular recreational beaches characterized by rip currents engage in various forms of public education and awareness strategies. The primary challenge facing beach safety practitioners engaged in these outreach programs is targeting a range of demographic groups including school children, adults, domestic and international tourists that have variable surf skills and knowledge, beach visitation rates, languages, and amounts of (dis)interest. Statistics show that most rip current drownings involve adolescent and adult males (Morgan et al., 2008; Gensini and Ashley, 2009; McCool et al., 2009; SLSA, 2009), but in practice it is extremely difficult to specifically target this demographic group. Rip current outreach programs have traditionally distributed generic brochures, stickers, refrigerator magnets, and posters, often with catchy slogans, to tourist information centers, holiday accommodations, and other tourist facilities near beaches with rip currents. The uptake

FIGURE 1.9 Cover of brochure for "Don't Get Sucked in by the Rip" pilot rip current intervention developed by the University of New South Wales. The campaign emphasized recognition of rip currents as dark gaps of "calmer" water between areas of breaking waves and whitewater. (*Source:* Williamson, A., Hatfield, J., Brander, R. et al. 2008. Improving beach safety: the Science of the Surf research project. *Proceedings of 2008 Australian Water Safety Conference*, Sydney, pp. 102–103. With permission.)

rate and effectiveness of these types of intervention material is not known although a study by Williamson et al. (2008) showed that brochures achieved the highest message recall (Figure 1.9).

Numerous beach safety programs deliver rip current education directly to primary and secondary school students. While these age groups rarely drown in rip currents, they represent a captive audience who can translate knowledge to their families, friends, and other children. School outreach programs are usually presented on site by professional lifeguards or volunteer lifesavers, government agencies, independent surf educators, and learn-to-surf schools. School programs involve a mixture of visual-based PowerPoint and DVD presentations with interactive discussions and explanations using various props. Based on USLA statistics, between 2005 and 2009, an average of 390,000 students per year received beach safety education by USLA lifeguards (USLA, 2009). In Australia, Surf Life Saving Australia estimates that it provides beach safety education to 240,000 students per year. Several programs in Australia involve students visiting a beach and floating in a rip current on a surfboard under careful supervision.

Beach and surf education is rarely compulsory for schools. A significant challenge common to most beach safety practitioners has been generating interest in surf education to principals and teachers in schools distant from coasts. While inland schools will argue that the information is not relevant to their students who "rarely go to the beach," the ignorance of these students about surf hazards places them in a high-risk category when they visit a beach with rip currents. Conversely, many schools located near beaches argue that surf education is redundant for children who have grown up along the coast. The reality is that all school children who will ever swim at a surf beach in their lifetimes will benefit from some type of beach and surf safety education. Until this type of program becomes a component of the basic school curriculum, even regionally, the number of students receiving information about rip currents will remain relatively small.

Providing rip current education to the adult population also presents a significant challenge. Aside from generic outreach material, other approaches have ranged from public service announcements played on community or hotel television channels to highway billboard advertising on approaches to popular beach destinations. In countries such as Australia, where beaches are major tourist attractions and rips are plentiful, most international tourists enter the country by air, yet little or no beach and rip safety information is made available to them during flight or upon arrival. However, Surf Life Saving Australia has recently arranged to provide surf education material in the form of an in-flight television commercial to several in-bound airlines during the most popular swimming months. Over two million inbound tourists will be reached by compulsory viewing with quarantine regulations prior to landing.

To date, little use has been made of popular Internet social communication networks such as YouTube, Facebook, and Twitter although the University of New South Wales and the Science of the Surf (SOS) program produced and posted an educational 4-minute rip current video titled "Don't Get Sucked in by the Rip" on YouTube in December 2008. This video surpassed 100,000 views by July 2010 and an improved version is now available free from UNSW TV (http://tv.unsw.edu.au).

NATIONAL RIP CURRENT EDUCATION CAMPAIGNS

Two main limitations apply to most existing rip education strategies, both globally and regionally. The first is the overall lack of funding devoted to the rip current problem. Even in the U.S. and Australia, beach safety practitioners struggle to attract significant funding devoted entirely to rip education. Compared to other natural hazards, the level of dedicated funding is exceedingly modest, putting a severe strain on resources and capabilities. The second limitation is the lack of consistent intervention strategies including materials, contents, and messages as well as an often ad hoc delivery approach. To overcome this problem, several countries have developed national rip education programs that focus outreach efforts around core standardized content so that the general public receives consistent safety information.

In the U.S., a collaborative development effort of USLA, NOAA's National Weather Service, and the National Sea Grant Program resulted in the implementation of the "Break the Grip of the Rip" campaign in 2005. This campaign is manifest by a standardized slogan and graphic (Figure 1.10a) that appears on generic beach signage, posters, stickers, DVD's, fact sheets, PowerPoint presentations, online games,

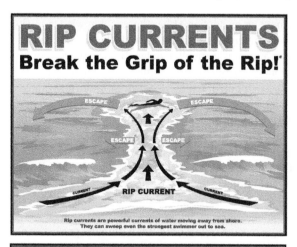

(a)

(b)

FIGURE 1.10 (a) Graphic used for "Break the Grip of the Rip" national rip current education program in the United States (www.ripcurrents.noaa.gov). (b) Graphic and core message of the Surf Life Saving Australia national rip education campaign launched in 2009 (*Source: Surf Life Saving Australia. 2009. National Coastal Safety Report.* With permission.)

and a travelling banner display for exhibitions and conferences (Davis and Painter, 2007). Most access points to U.S. public beaches now have "Break the Grip of the Rip" safety signs. Much of the other material is mass produced and made available on the campaign's website (www.ripcurrents.noaa.gov). In addition, NOAA has designated the first full week of June preceding the summer season as "Rip Current Awareness Week."

However, without an evaluation of beachgoers' knowledge of rip currents before the launch of the campaign and problems with the accuracy of incident reporting of rip current drowning, it is difficult to judge the effectiveness of "Break the Grip of the Rip." Anecdotally, increased use of the "rip current" term by both the media and the public has been noted, but there has been no formal evaluation of how well the campaign outreach material has translated into an increased understanding of the rip current hazard by the beach going public and a determination of which forms of the outreach program have been the most effective.

Surf Life Saving Australia launched a 3-year campaign in 2009 to educate all Australians about rip currents. The campaign was based initially on the slogan, "To Escape a Rip, Swim Parallel to the Beach," an accompanying website containing educational material (www.ripcurrents.com.au), and widespread advertising including public service announcements on major radio and television networks (Figure 1.10b). While the campaign received some criticism within the professional ocean lifeguard community and from independent surf educators, largely due to disagreement with the appropriateness of the core slogan, the media attention created by this controversy generated significant public interest in and attention to the rip current problem. Results from a Newspoll evaluation of the campaign by SLSA revealed significant recall of the slogan message. Like all campaigns however, it is impossible to assess significant positive impacts on the incidence of rip current rescues and drownings for many years to come.

CONVENTIONAL RIP CURRENT ADVICE: SWIM PARALLEL OR STAY AFLOAT?

A review of beach safety websites around the world will invariably lead to advice on how to respond when caught in a rip current that generally consists of variations of four common points: (1) do not panic; (2) do not swim against the rip; (3) escape the rip by swimming parallel to the beach; or (4) stay afloat and signal for help. While there is agreement about the importance of (1) and (2), considerable debate now exists over the appropriate order of (3) and (4), largely due to recent advances in scientific measurements and observations of rip current circulation described by MacMahan et al. (2010).

The "swim parallel to the beach" advice for escaping a rip has existed for well over 50 years and is largely related to the traditional paradigm of rip current flow shown in Figure 1.2. The advantages of this behavioral response are straightforward: (1) it gets a swimmer out of the rip and onto the safer confines of a shallow sand bar before he or she is far offshore; (2) as rips are narrow, it is much easier to swim to the side of a rip current than against the direction of rip flow back to shore; and (3) this advice is often the only self-escape option available to swimmers on unpatrolled beaches in the absence of lifeguards or other help. The flaws with this advice are:

1. It is not an option for people who cannot swim.
2. Many rips flow at extreme angles to the beach and a swimmer may end up actually swimming against the rip flow.
3. Many rips are characterized by strong flows of water draining off the sides of sandbars and/or alongshore drift across the bars associated with the rip cell circulation, both of which may prove difficult to swim against.
4. A swimmer is still carried offshore as he or she swims to the side of the rip.
5. The physical exertion required may contribute to exhaustion and panic if the swimmer does not successfully escape the rip.

To overcome the problem with rip geometry and the use of the "parallel" term, some education programs encourage swimmers to swim toward the "side of the rip" or "toward whitewater and breaking waves," but this assumes knowledge of rip identification and does not resolve the other disadvantages associated with this strategy. Swimming parallel to the beach is best suited for strong, confident swimmers and swimmers who find themselves seaward of the breakers—swimming parallel to the beach and then back to shore across sand bars where the waves are breaking can bring them back to the beach.

In contrast, "staying afloat" in a rip is a passive behavioral response that ideally involves remaining calm and signaling for help. This advice is supported by the research of MacMahan et al. (2010) indicating that rip current circulations are contained mostly within the surf zone and will bring drifting objects back to the relative safety of shallow sand bars within minutes. In the presence of lifeguards, surfers, or other help, staying afloat is certainly a good option, particularly for weak swimmers and non-swimmers. However, MacMahan et al. (2010) also cited a 20% chance of exiting the surf zone from a rip current. Furthermore, rip channels are rarely completely free of breaking waves. For an inexperienced surf swimmer or a non-swimmer told to stay afloat, the combination of waves breaking overhead while being carried quickly offshore could contribute to panic. On an unpatrolled, sparsely populated, or remote beach, the consequences may be fatal.

This debate has no winner. Rip current flow is non-uniform and unsteady and can vary significantly with tidal elevations and among different types of rips. For this reason, the efficacy and appropriateness of the "swim parallel" and "stay afloat" approaches can be variable and inconsistent. Both can work and both can be seriously flawed. In the absence of an overriding satisfactory behavioral response for beachgoers caught in rips, the best advice to prevent rip current drownings should be preventative: stop people from entering rips in the first place. This means providing swimmers with more motivation to swim in supervised areas or improving beachgoers' abilities to understand and recognize rip current hazard.

PUBLIC AWARENESS OF RIP CURRENTS

A common statistic used to describe rip current hazard in the U.S. is that in an average year, rip currents are responsible for more deaths than hurricanes, tornados, floods, lightning strikes, or sharks (Lushine, 1991; Fletemeyer and Leatherman, 2010). A similar argument could be made for Australia with the addition of bushfire hazards. Nevertheless, rip currents, which are relatively high frequency, low magnitude hazards, are largely overlooked by governments and coastal managers in comparison to more infrequent, but often catastrophic, natural hazards. As described by Short and Hogan (1994), this is based on several reasons, for example, (1) rips generally do not pose obvious threats to structures or property, (2) rips cannot be managed via engineering devices, and (3) the impacts of rip drownings and injuries are largely restricted to a victim's immediate family and the authorities (lifeguards, emergency services, hospitals, and coroners) dealing with the incidents.

Another element is complacency: local, state, and federal governments assume that lifeguards and warning signs are sufficient to keep beachgoers safe. Unfortunately, this complacency has contributed to a high degree of ignorance among the beachgoing public.

The reality is that most beachgoers still do not understand what rip currents are or know how to identify them. A recent study by Williamson et al. (2008) found that almost 80% of surveyed Australian beachgoers were aware of common rip safety advice such as "swim parallel to the beach," but only 40% could identify a rip current when shown a picture of one, even though 80% thought they could. Of more concern, half the respondents indicated that rip currents were the safest places to swim. In a study of the beach safety knowledge of international and domestic students on beaches in Queensland, Australia, Ballantyne et al. (2005) found similar results. Of all the students in both groups who claimed to know what a rip was, two-thirds said they did not know how to recognize one. Only 18% said that rip currents could be identified by the presence of apparently calm water between breaking waves. When shown a photograph of a beach with two fixed rip currents, 61% of all students selected them as places where they would swim. This is particularly disturbing because fixed rip currents are the most common types along the east coast of Australia and their most common visual indicator is a darker, seemingly calm area of water between breaking waves (Figures 1.4 and 1.5). Based on these results, it is clear that one of the fundamental knowledge disconnects is the inability of beachgoers to visually identify rip currents. As highlighted by Ballantyne et al. (2005), basic awareness of rips and their potential dangers is worthless without an ability to recognize them.

CHALLENGES FOR RIP CURRENT RESEARCH

Scientific research on rip currents is ongoing despite the logistical challenges and demands of obtaining field measurements in a demanding environment. As described previously, we now have a solid, basic understanding of rip flow behavior, but existing measurements have almost exclusively been made in open beach fixed rip current systems. More work must be done to examine the Lagrangian and Eulerian flow behaviors for different types of rips such as topographic and flash rips under differing wave energy and tidal regimes. Much of this research will have direct applications to beach safety. A strong need also exists to examine the role of rip current type, tidal elevation and flow pulsing on the incidence of rip current rescues. The use of Lagrangian drifters and GPS can also be extended to the use of human rip floaters (Short and Hogan, 1994; Brander and Short, 2000) to directly assess the efficacy of the "swim parallel" and "stay afloat" behavioral responses for swimmers caught in rips.

The primary issue, however, is the paucity of studies relating to the social and psychological aspects of rip current science. Very little information exists regarding the demographics, behavior, and fundamental rip and beach safety knowledge of beachgoers. In particular, little is known about choice of swim locations, victims rescued from rips, and what people perceive and how they respond when caught in rips. Rip currents do not drown people—panic does—and the factors contributing to the onset of panic in swimmers caught in rips are poorly understood. Until social and behavioral science techniques are applied to rip current research to determine what people know or can learn about rip currents, the disconnects among rip current science, beach safety practitioners, and beachgoers (Figure 1.1) will remain.

CHALLENGES FOR OUTREACH

Existing rip current outreach strategies are clearly successful. For example, since 1960 in the United States, both beach attendance and the incidence of surf rescues have risen, but the total number of drownings has remained relatively stable (Branche and Stewart, 2001). Nevertheless, rip current drownings continue and interest in improving rip current outreach programs in developed countries such as the U.S., Australia, and the United Kingdom has never been greater. However, while the establishment of national rip education campaigns is a major step forward, a huge challenge for beach safety practitioners is evaluating the effectiveness of these programs in a meaningful way. Are the strategies having a positive impact on beachgoers' knowledge and behavior with respect to rip current hazard? Which approaches and types of intervention media are the most successful? What do people remember about rip current education? Until educational programs are evaluated and modified to suit the findings of social and psychological rip current research, the incidence of rip current drownings will likely not improve.

Effective rip current education strategies must extend beyond slogans and catch phrases and focus on existing knowledge gaps about rip currents. The ultimate challenge for rip current outreach is to improve beachgoer awareness and understanding of rip current hazard so that they are increasingly motivated to seek out and swim only at beaches supervised and patrolled by lifeguards. However, because people will always swim outside patrolled areas, the best alternative is to promote rip avoidance and improve rip identification skills. The primary reason beachgoers are deficient in identifying rips is that few efforts have been dedicated to teaching them how to spot rips. Rips are very visible features and education strategies should focus on the most common type that is both the easiest to identify and is also responsible for most rescues and drownings: fixed rip currents that appear as darker gaps between areas of whitewater.

For a domestic population, the only long-term solution to ingrain rip identification in beach cultures is to teach primary and high school children basic rip current knowledge. In the short term, the greatest challenges for outreach programs are the domestic and international tourist populations. Tourists associate beaches with fun and pleasant activities, are more likely to partake in risky behavior, and are not

(a)

(b)

FIGURE 1.11   (*See color insert.*) Time sequence of release of purple dye into a topographic rip current during a public demonstration at Tamarama Beach, Sydney, Australia. (Courtesy of Robert W. Brander.)

(c)

FIGURE 1.11   (continued).

interested in being educated while on holiday (Ballantyne et al., 2005). The tourism industry at large is also wary of promoting negative content associated with a major recreational amenity.

Approaches to rip current outreach must therefore be as visual and engaging as possible, utilizing easily accessible internet social media and technology such as Facebook, YouTube, and Twitter. The use of video footage on these media channels is a powerful short-term educational strategy. Figure 1.11 shows a sequence of colored dye in a rip current rapidly heading offshore. The dye is a powerful, quick, eye-catching and non-threatening tool to illustrate and increase awareness of the nature of the rip current hazard. Preliminary data from Australia indicate that the release of visible dye is a particularly effective educational method.

Other challenges facing future rip current outreach include improving the type, consistency, and accuracy of data associated with incident reporting. Lifeguards are ideally placed as data collectors because they are on-site for extended periods to assess beach conditions, estimate crowds, and conduct rescues. However, previous studies (Williamson, 2006) have shown that the reliability of lifeguards as data collectors on the beach can be extremely variable. Nevertheless, the spatial and temporal coverage of beaches by lifeguards to gather data cannot be replicated by independent researchers and should become an established job responsibility and requirement. Furthermore, lifeguards have opportunities to be more proactive in conducting educational rip demonstrations on beaches as time and conditions permit (Klein et al., 2003).

Finally, a need exists to dramatically improve rip current safety in many developing countries. While this initially involves the establishment of trained lifeguard services, rip education has positive results. Klein et al (2003) found that most surf accidents in a popular beach region in Brazil occurred where warning flags were clearly displayed, suggesting a lack of knowledge of or respect for the warning system. A safety education campaign utilizing videos, newspapers, leaflets, and signs was developed and an 80% reduction in the number of fatal accidents in the region was reported several years after implementation. The campaign materials were based on social, economic, and cultural data collected by beach lifeguards.

CONCLUSIONS

This chapter has attempted to describe the information disconnects that have hindered efforts to reduce the incidence of rip current drowning. The next decade presents a critical opportunity to overcome these disconnects as interest in rip currents from all sectors has never been higher. More countries with beach drowning problems are seeking to improve existing outreach strategies. This is evident from the wide range of participants at the First International Rip Current Symposium that provided an important opportunity for rip current scientists and beach safety practitioners to interact.

The first step to overcoming these disconnects is to encourage open two-way communications and collaborations between rip current scientists and beach safety practitioners. The second step is to focus on the beachgoers. What do we know about beachgoers? What do they know about rip currents? Social and behavioral sciences must investigate beachgoer demographics, their choices of swim locations, swimming abilities and behaviors, knowledge of surf hazards, and responses of swimmers caught in rips. This information is vital in order to develop and tailor appropriate preventative rip current education strategies.

Finally, it is important to remember that rip currents are not dangerous to beach users who understand and recognize them. Experienced swimmers, surfers, and lifeguards actively seek out and use rip currents to their advantage. Informed beachgoers can make correct decisions about where to swim in the presence or absence of lifeguards based on their knowledge of rip currents. The lack of understanding is the only factor that makes rip currents dangerous and this lack can be overcome only by improving beachgoer awareness of rip currents and their ability to identify the currents. A person will not drown in a rip current if he or she does not swim in one. The development of successful rip current science and outreach will take time, but most importantly, will require significantly more funding than is presently devoted to such efforts.

REFERENCES

Aagaard, T., Greenwood, B., and Nielsen, J. 1997. Mean currents and sediment transport in a rip channel. *Marine Geology*, 140: 25–45.
Aagaard, T. and Vinther, N. 2008. Cross-shore currents in the surf zone: rips or undertow? *Journal of Coastal Research*, 24: 561–570.
Austin, M., Scott, T., Brown, J. et al. 2010. Temporal observations of rip current circulation on a macro-tidal beach. *Continental Shelf Research*, 30: 1149–1165.
Ballantyne, R., Carr, N., and Hughes, K. 2005. Between the flags: an assessment of domestic and international university students' knowledge of beach safety in Australia. *Tourism Management*, 26: 617–622.

Bech, E. 2007. Rip currents: a public awareness negligence. *Proceedings of World Conference on Drowning Prevention*, Matosinhos, Portugal. International Life Saving Federation, http://drowningprevention.ilsf.org/2007/papers.

Bowen, A.J. 1969. Rip currents 1. Theoretical investigations. *Journal of Geophysical Research*, 74: 5467–5478.

Bowman, D., Rosen, D.S., Kit, E. et al. 1988. Flow characteristics at the rip current neck under low energy conditions. *Marine Geology*, 79: 41–54.

Branche, C.M. and Stewart, S., Eds. 2001. *Lifeguard Effectiveness: A Report of the Working Group*. Centers for Disease Control and Prevention, Atlanta.

Brander, R.W. and Short, A.D. 2000. Morphodynamics of a large-scale rip current system at Muriwai Beach, New Zealand. *Marine Geology*, 165: 27–39.

Brander, R.W. and Short, A.D. 2001. Flow kinematics of low-energy rip current systems. *Journal of Coastal Research*, 17: 468–481.

Carey, W. and Rogers, S. 2005. Rip currents: coordinating coastal research, outreach and forecast methodologies to improve public safety. *Proceedings of Coastal Disasters*, ASCE, New York, pp. 285–296.

Chen, Q., Dalrymple, R.A., Kirby, J.T. et al. 1999. Boussinesq modeling of a rip current system. *Journal of Geophysical Research*, 104: 20617–20638.

Dalrymple, R.A. 1978. Rip currents and their causes. *Proceedings of 16th International Conference on Coastal Engineering*, ASCE, New York, pp. 1474–1427.

Davis, W.M. 1925. Undertow. *Science*, 62: 33.

Davis, R.A, Jr. and Fitzgerald, D.M. 2004. *Beaches and Coasts*. Blackwell, Oxford.

Davis, P. and Painter, A. 2007. NOAA/USLA *rip current awareness campaign*. Proceedings of the World Conference on Drowning Prevention, Matosinhos, Portugal. International Life Saving Federation, http://drowningprevention.ilsf.org/2007/papers.

Dean, R.G. and Dalrymple, R.A. 2002. *Coastal Processes with Engineering Applications*. Cambridge University Press, New York.

Engle, J., MacMahan, J., Thieke, R.J. et al. 2002. Formulation of a rip current predictive index using rescue data. *Proceedings of National Conference on Beach Preservation Technology*, FSBPA, Biloxi, MS.

Fletemeyer, J. and Leatherman, S. 2010. Rip currents and beach safety education. *Journal of Coastal Research*, 26: 1–3.

Fowler, R.E. and Dalrymple, 1990. Wave group forced nearshore circulation. *Proceedings of 22nd International Conference on Coastal Engineering*, Delft, The Netherlands, pp. 729–742.

Garcia-Faria, A.F., Thornton, E.B., Lippmann, T.C. et al. 2000. Undertow over a barred beach. *Journal of Geophysical Research*, 105: 16999–17010.

Gensini, V.A. and Ashley, W.S. 2009. An examination of rip current fatalities in the United States. *Natural Hazards*, 54: 159–175.

George, P. 2007. International signs and beach safety flags: is it possible to achieve an international beach safety flag system? *Proceedings of World Conference on Drowning Prevention*, Matosinhos, Portugal. International Life Saving Federation, http://drowningprevention.ilsf.org/2007/papers.

Haas, K.A. and Svendsen, I.A., 2002. Laboratory measurements of the vertical structure of rip currents. *Journal of Geophysical Research*, 107: 3047; doi: 10.1029/2001JC000911.

Hartmann, D. 2006. Drowning and beach safety management (BSM) along the Mediterranean beaches of Israel: a long-term perspective. *Journal of Coastal Research*, 22: 1505–1514.

Holman, R.A., Symonds, G., Thornton, E.B. et al. 2006. Rip spacing on an embayed beach. *Journal of Geophysical Research*, 111; doi:10.1029/2005/JC002965.

Inman, D.L. and Quinn, W.H. 1952. Currents in the surf zone. *Proceedings of 2nd Conference on Coastal Engineering*. ASCE, Houston, pp. 24–36.

Inman, D.L. and Brush, B.M. 1973. Coastal challenge. *Science*, 181: 20–32.

Johnson, D. and Pattiaratchi, C. 2004. Transient rip currents and nearshore circulation on a swell-dominated beach. *Journal of Geophysical Research*, 109; doi. 10.1029/2003JC001798.

Klein, A.H da F., Santana, G.G., Diehl, F.L. et al. 2003. Analysis of hazards associated with sea bathing: results of five years' work in oceanic beaches of Santa Catarina State, Southern Brazil. *Journal of Coastal Research*, SI 35: 107–116.

Komar, P.D. 1998. *Beach Processes and Sedimentation*, 2nd ed. Prentice Hall, New York.

Lascody, R.L. 1998. East Central Florida rip current program. *National Weather Digest*, 22: 25–30.

Lippmann, T.C and Holman, R.A. 1989. Quantification of sand bar morphology: a video technique based on wave dissipation. *Journal of Geophysical Research*, 94: 995–1011.

Longuet-Higgins, M.S. and Stewart, R.W. 1964. Radiation stress in water waves: a physical discussion with applications. *Deep-Sea Research*, 11: 529–563.

Lushine, J.B. 1991. A study of rip current drownings and weather related factors. *National Weather Digest*, 16: 13–19.

MacMahan, J., Brown, J., Thornton, E. et al. 2010. Mean Lagrangian flow behavior on an open coast rip-channeled beach: a new perspective. *Marine Geology*, 268: 1–15.

MacMahan, J.H., Brown, J., and Thornton, E. 2009. Low-cost hand-held Global Positioning System for measuring surf-zone currents. *Journal of Coastal Research*, 25: 744–754.

MacMahan, J.H., Thornton, E.B., and Reniers, A.J.H.M. 2006. Rip current review. *Coastal Engineering*, 53: 191–208.

MacMahan, J.H., Reniers, A.J.H.M., Thornton, E.B. et al. 2004. Infragravity rip pulsations. *Journal of Geophysical Research*, 109; doi: 10.1029/2003JC002068.

MacMahan, J.H., Thornton, E.B., Reniers, A.J.H.M. et al. 2008. Low-energy rip currents associated with small bathymetric variations. *Marine Geology*, 255: 156–164.

MacMahan, J.H., Thornton, E.B., Stanton, T.P. et al. 2005. RIPEX: observations of a rip current system. *Marine Geology*, 218: 118–134.

Masselink, G. and Hughes, M.G. 2003. *Introduction to Coastal Processes and Geomorphology*. Oxford University Press, Oxford.

McCool, J.P., Ameratunga, S., Moran, K. and Robinson, E. 2009. Taking a risk perception approach to improving beach swimming safety. *International Journal of Behavioural Medicine*. doi 10.1007/s12529-009-9042-8.

McCool, J.P., Moran, K., Ameratunga, S. et al. 2008. New Zealand beachgoers' swimming behaviours, swimming abilities, and perception of drowning risk. *International Journal of Aquatic Research and Education*, 1: 7–15.

Morgan, D., Ozanne-Smith, J., and Triggs, T. 2008. Descriptive epidemiology of drowning deaths in a surf beach swimmer and surfer population. *Injury Prevention*, 14: 62–65.

Pattiaratchi, C., Olsson, D., Hetzel, Y. et al. 2009. Wave-driven circulation patterns in the lee of groynes. *Continental Shelf Research*, 29: 1961–1974.

Ranasinghe, R., Symonds, G., Black, K. et al. 2000. Processes governing rip spacing, persistence and strength in a swell dominated, microtidal environment. *Proceedings of 27th International Conference on Coastal Engineering*, Sydney, ASCE, pp. 455–467.

Ranasinghe, R., Symonds, G., and Holman, R.A. 2004. Morphodynamics of intermediate beaches: a video imaging and numerical modelling study. *Coastal Engineering*, 51: 629–655.

Reniers, A., MacMahan, J., Thornton, E.B. et al. 2009. Surf zone retention on a rip channelled beach. *Journal of Geophysical Research*, 114, C10010; doi:10.1029/2008JC005153.

Scott, T.M., Russell, P.E., Masselink, G. et al. 2009. Rip current variability and hazard along a macro-tidal coast. *Journal of Coastal Research*, SI50: 1–6.

Shepard, F.P. 1936. Undertow: rip tide or rip current? *Science*, 84: 181–182.

Shepard, F.P., Emery, K.O., and Lafond, E.C. 1941. Rip currents: a process of geological importance. *Journal of Geology*, 49: 338–369.

Shepard, F.P. and Inman, D.L. 1950. Nearshore circulation. *Proceedings of First Conference on Coastal Engineering*, ASCE, New York, pp. 50–59.

Shepard, F.P. and Inman, D.L. 1951. Nearshore circulation related to bottom topography and wave refraction. *Transactions American Geophysical Union*, 31(4): 196-213.

Sherker, S., Brander, R.W., Finch, C. et al. 2008. Why Australia needs an effective national campaign to reduce coastal drowning. *Journal of Science and Medicine in Sport*, 11: 81–83.

Sherker, S., Williamson, A., Hatfield, J., et al. 2010. Beachgoers' beliefs and behaviours in relation to beach flags and rip currents. *Accident Analysis and Prevention*, 42: 1785–1804.

Short, A.D. 2007. Australian rip systems: friend or foe? *Journal of Coastal Research,* SI50: 7–11.

Short, A.D. and Brander, R.W. 1999. Regional variations in rip density. *Journal of Coastal Research*, 15: 813–822.

Short, A.D. and Hogan, C.L. 1994. Rip currents and beach hazards: their impact on public safety and implications for coastal management. *Journal of Coastal Research*, SI12: 197–209.

Sonu, C.J. 1972. Field observations of a nearshore circulation and meandering currents. *Journal of Geophysical Research*, 77: 3232–3247.

Surf Life Saving Australia. 2009. *National Coastal Safety Report.*

Tang, E.C.S. and Dalrymple, R.A. 1989. Nearshore circulation: rip currents and wave groups. *Nearshore Sediment Transport*, Seymour, R.J., Ed. Plenum Press, New York.

Thornton, E.B., Sallenger, A.H., and MacMahan, J.H. 2007. Rip currents, cuspate shorelines, and eroding dunes. *Marine Geology*, 240: 151–167.

Turner, I.L., Whyte, D., Ruessink, B.G. et al. 2007. Observations of rip spacing, persistence and mobility at a long, straight coastline. *Marine Geology*, 236: 209–221.

Williamson, A. 2006. Feasibility study of a water safety data collection for beaches. *Journal of Science and Medicine in Sport*, 9: 243–248.

Williamson, A., Hatfield, J., Brander, R. et al. 2008. Improving beach safety: the Science of the Surf research project. *Proceedings of 2008 Australian Water Safety Conference*, Sydney, pp. 102–103.

United States Lifesaving Association. 2009. *National Lifesaving Statistics*, http://www.usla.org/Statistics/public.asp.

2 Flash Rip Currents on Ocean Shoreline of Long Island, New York

Michael P. Slattery, Henry Bokuniewicz, and Paul Gayes

CONTENTS

INTRODUCTION

Rip currents and their inherent dangers have long served as topics for both scientific research and public discussion (MacMahan et al., 2006; Murray et al., 2003; Hammack et al., 1991; Dalrymple and Lozano, 1978; Bowen and Inman, 1969). Despite the long history of observations that led to an in-depth understanding of this phenomenon, predictive models that accurately describe conditions favorable for rip current development have been elusive. Part of the difficulty is that rip currents can be produced by a variety of mechanisms (MacMahan et al., 2006) and exhibit a wide range of characteristics. Four types are commonly recognized: (1) fixed, (2) permanent, (3) traveling, and (4) flash rip currents (Brewster, 1995).

Fixed and permanent rip currents are controlled by coastal morphology although their existence and intensity varies with conditions of waves and tides. Permanent rip currents are associated with immobile features of a rocky shoreline or permanent coastal structures like groins. Fixed rips, on the other hand, are associated with undulation in morphology on a sand beach that can change location and intensity more or less quickly over time.

Rip currents are still common, however, in the absence of favorable morphology, although their appearance and strength may be different. Rip currents observed along straight portions of a barless shoreline tend to appear as single, shore-perpendicular lines of foam (Murray et al., 2003). These may be flash rip currents or traveling types. Flash rip currents usually appear weaker than their bathymetrically controlled counterparts and last only a few minutes. Flash rips are controlled by nearshore oceanographic processes and do not rely on the influence of bathymetric undulations. While the "flash rip" term is fairly recent, theories regarding their generation go back decades. Bowen and Inman (1969) described nearshore traits related to incident waves that can influence water pressure and induce rip currents. Traveling rip currents, like flash rip currents, also owe their existence to wave, wind, and tide conditions and are not dependent on undulations in beach morphology. Traveling rip currents, as their name implies, move along a shoreline in the direction of the prevailing waves.

The factors that lead to initial formation of a rip may be obscure (Brander and Short, 2000; Aagard et al., 1997), but it seems reasonable that flash rips caused by oceanographic processes appear first. Alongshore modulation of breaker heights due to wave refraction or synchronous edge waves has been proposed to initiate rip currents (Komar, 1998). Wave refraction or infragravity waves are generated by variable heights in wave sets and may generate the required difference in longshore hydraulic heads. On reflective beaches, a longshore edge wave may develop from long period waves superimposed upon the surf (Murray et al., 2003). Long period waves reflecting off a beach, headland, or hard structure alongshore can produce resonance within the surf zone, creating a standing edge wave (Komar, 1998).

The isolated seaward current may increase energy dissipation in the surf zone, decreasing wave heights locally and in turn accentuating variations in longshore set-up and feeding back into a stronger rip current (Murray et al., 2003). If the rip current persists long enough on a sandy beach, shoreline undulation or gaps in the bar may develop, extending the lifetime and strength of the seaward flow. Flash rip currents thus uncomplicated by coastal morphology may provide clues about the factors initiating rip current generation.

The purpose of this study was to characterize the rip current regime along the oceanic south shore of Long Island, New York. The occurrences of flash rip currents were documented, and the potential for infragravity waves as a cause of rip currents was investigated.

STUDY AREA

Long Island's ocean coast is relatively straight (Figure 2.1). Average wave conditions are 1 m in height, periods of 7 sec, generally approaching from the southeast, although nor'easter storms and an occasional hurricane or tropical storm may elevate wave heights above 3 m and extend periods near 14 sec (Buonaiuto, 2003). The tidal range is fairly limited—below 1 m even during spring tides. Wave approach drives an average longshore sand transport from east to west (Kana, 1995), but reversals are common. Infragravity waves observed along the shoreline (Schubert and Bokuniewicz, 1991) revealed periods between 20 and 300 sec, with a "pronounced, longer period oscillation outside the traditional infragravity band with a period of about 1000 sec."

FIGURE 2.1 Study locations along the south shore of Long Island, New York are defined by a long straight coastline.

Rip currents (sometimes called "sea pusses" locally) are known to occur on the south shore of Long Island and constitute hazards to beachgoers. Based on the number of rescues reported by lifeguards, the 2009 season appeared unusually active (Bleyer, 2009). Rip current drownings accounted for 19 deaths from 2008 to 2010 (Nelson Vaz, National Weather Service, personal communication). The fatalities were clustered (87% of 2008 drownings occurring over 3 days; 80% of 2009 drownings over 4 days; and 66% of 2010 drownings within a week). Interestingly, only two females were among the fatalities and both were under the age of 13.

Two sites were examined for this study—one at East Hampton and the other at the Fire Island Lighthouse (Figure 2.1). The beach is fairly steep (+1.5 m to –1.5 m) and still water at both locations (0.04 at East Hampton and 0.06 at Fire Island) makes the beaches reflective. In addition, each location has an ephemeral bar located 227 m offshore in a water depth of 3 m at the bar crest (Figure 2.2).

METHODS

Rip currents are typically apparent to a casual observer as a zone of turbid, churned water extending through the breakers (Turner et al., 2007; MacMahan et al., 2005; Komar, 1998). An Erdman camera was installed in February 2007 atop a house perched on a 9-m dune in East Hampton for a total elevation of approximately 15 m above mean sea level and 120 m from the shoreline. A series of three groins were

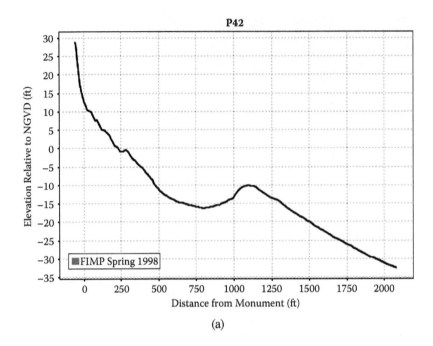

(a)

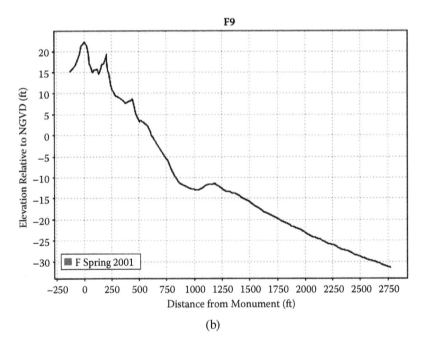

(b)

FIGURE 2.2   Both beaches are steep and have ephemeral bars that can form at a position 330 m from the dune crest (227 m offshore) at (a) East Hampton and (b) Fire Island.

located approximately 1 km west of the camera. The groins extend approximately 100 m out from the dune and 23 m into the surf zone. These structures and the resulting shoals formed an artificial headland along an otherwise straight section of coastline (Bokuniewicz, 2004).

At East Hampton, the camera faces shore-normal for 9 hr during daylight although sun glare necessitated a 5-hour period looking alongshore. The observable beach from the two camera orientations is approximately 1 km, although the shore-normal view only covers 120 m of the surf zone. Images are recorded every 15 sec for 14 hr, but seasonal shifts in sunlight cause a variability of ±3 hr of usable imagery. At Fire Island, a second camera was installed in the lens room at the top of the lighthouse, approximately 50 m above sea level and 350 m from the water in April 2009.

Images were examined visually for the presence of rip currents based on key traits, particularly lines of foam or lack of wave breaks, as is common for this type of video monitoring study (Holman et al., 2006; Murray et al., 2003). For East Hampton, 171,443 images were reviewed, accounting for over 714 hr of shore observations. At the Fire Island Lighthouse, 155,677 images covering 865 hr of beach observations have been analyzed. Average, minimum, and maximum rip duration, average number of rips per day per kilometer of beach, and rip frequency as a percentage of total time observed were tabulated. Rip activity (RA) was calculated by normalizing the time rips present to the total time and distance observed along the shore (Murray et al., 2003).

Evidence of very long period infragravity waves (Schubert and Bokuniewicz, 1991) pointed to edge waves as possible causes of rip currents. Recording long period waves is problematic because they are not detectable visually and instrumentation deployment in a surf zone is cumbersome and expensive. Cables often snap, and equipment may be lost due to rapid burial. Therefore, land-based video monitoring is a key component of many rip studies. Seismometers were also used to measure wave conditions at East Hampton.

Seismic records used to detect earthquakes have always included considerable background noise (Hasselman, 1963). The microseisms noted for near-shore stations were associated with ocean noise (Bromirski and Dunnebier, 2002; Darbyshire and Okeke, 1969; Hasselman, 1963) and revealed considerable amounts of energy at the longer periods (Ruessink, 1998; Oliver and Ewing, 1957). Periodicities in the seismic signals have been used to trace surf noise—one at the incident wave period and another approximately double the period due to standing infragravity waves (Oliver and Ewing, 1957). In addition, seismic records were obtained at East Hampton to monitor tidal influence on coastal groundwater (West and Meke, 2000).

A seismometer was established in May 2007 at the East Hampton site on a solid concrete floor about 120 m from the shoreline. The seismometer recorded 20 samples per second. Spectra were produced from selected 10-hr segments of data to identify dominant frequencies recorded at the study location. In April 2009, a second seismometer was established approximately 3 km east of the main study location at the Maidstone Club (Figure 2.3). The nearest source of open-ocean data is NOAA buoy 44017 located 31 km offshore. Recorded wave periods from the buoy were correlated to the dominant peaks recorded by the seismometers.

FIGURE 2.3   Second seismometer deployed at Maidstone Club, approximately 2.3 km east of the original East Hampton study site.

RESULTS

Images from East Hampton (Figure 2.4) were used to detect 49 distinct rip events spanning 241 scenes (Table 2.2). The rips were generally narrow, on the order of 15 m, and did not extend more than 31 m offshore. The average duration of events was approximately 76 sec (median = 60 sec; Table 2.1). Maximum rip duration was 150 sec, observed on four occasions (Figure 2.5). Thirty of the rip currents lasted between 45 and 75 sec, accounting for 61% of the total rip currents observed. Of the 714 hr covered by the images, 0.14% displayed the presence of rip currents with no observed preference for time of day or position in the scenes. Conditions generated an average of two rips per kilometer per day (Table 2.2). Initial results show a preference for rip currents near the time of high water for East Hampton.

At Fire Island, 130 distinct rip events spanning 338 scenes were observed (Figure 2.6, Table 2.2). The rip currents at this location were also narrow (e.g., 15-m widths) and extended only 30 m offshore. Average duration of these rip currents was 52 sec with a median of 40 sec. Eighty-eight of these rips lasted between 30 and 60 sec (Figure 2.5), accounting for 68% of the total. Of the 865 hr covered, 0.22% displayed the presence of rip currents with no preferred time of day, and there seemed to be only a very slight preference for rip formation to the east side of the viewable area (Table 2.1). For this stretch of beach, 13 rips per kilometer per day can be expected. The data obtained plot at nearly zero for rip activity (Figure 2.7) on the graph by Murray et al. (2003).

FIGURE 2.4   Sample rip current similar to those observed off East Hampton.

TABLE 2.1
Rip Duration and Activity

	Fire Island	East Hampton
Time observed	865	714
Average rip duration (sec)	52	76
Minimum duration (sec)	20	30
Maximum duration (sec)	180	150
Rip activity	4.65E-04	1.44E-06

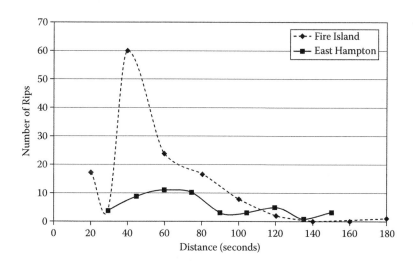

FIGURE 2.5   Rip durations for East Hampton and Fire Island.

TABLE 2.2
Rip Frequencies and Occurrences

	Fire Island	East Hampton
Scenes observed	155,677	171,443
Scenes with rip currents	338	241
% Frequency	0.22	0.14
Distinct rip events	130	49
Rips/km/day	13	2

FIGURE 2.6   Sample rip current commonly observed off Fire Island.

Seismic records from the original study location demonstrated spectral peaks at a range of frequencies (Figure 2.8). Most peaks fell between 0.33 Hz (3 sec) and 0.067 Hz (15 sec) and less than 0.05 Hz (20 sec) (Table 2.3). Seventeen peaks were detected in the spectrum at the first site, and 12 were evident from a second nearby location in East Hampton. When viewed with respect to the measured wave periods at NOAA buoy 44017, a corresponding period was observed in the spectra for all but one day.

DISCUSSION

The slope of the beach has been hypothesized to cause (1) an increased prevalence of flash rip currents on steeper beaches if they are associated with offshore-directed flows of meandering longshore currents (Smith and Largier, 1995) or (2) a decreased prevalence on steeper beaches where the surf zone is narrow and the water flux is reduced (Murray et al., 2003). Further findings from Murray suggested that rip

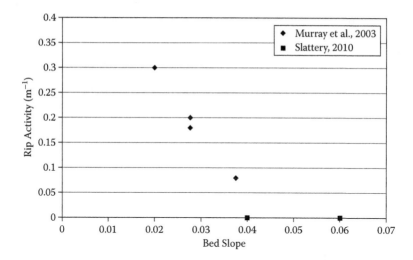

FIGURE 2.7 Rip activity (RA) and beach slope from this study and that of Murray et al. (2003).

activity (RA) becomes effectively zero at slopes >0.03—a conclusion supported by our findings at both East Hampton where the slope was 0.04 (RA = 1.44 * 10 – 6) and at Fire Island where the slope was 0.06 (RA = 4.0 * 10 – 4).

This trend is more apparent when our results are plotted in conjunction with measured results from Murray et al. (2003). While the result was not zero, the evidence from Long Island supports their claim that RA is essentially zero at slopes >0.03. Rip currents along the Long Island south shore would therefore be likely candidates for edge wave formation and less likely to be dominated by the position and morphology of the offshore bar. Further indications that support these findings are the narrowness of the rips and the lack of preferential location of formation—both characteristics that make bar control unlikely (Dalrymple and Lozano, 1978; Bowen and Inman, 1969).

Rip current positions were widely variable within the scenes. While Fire Island rip currents showed a slight preference for a location to the east of the viewable area, they apparently were not prevalent long enough to modify the beach morphology. An oddity to this study that may further support a detachment from bathymetric control was the slight prevalence of rip currents within 3 hr before and after high tide—the reverse of the typical case. Infragravity waves were detected seismically, but an association with rip current occurrence was inconclusive. Spectral peaks at the frequencies indicating infragravity waves (<0.05 Hz) were common, but not ubiquitous in recordings of the East Hampton seismic stations. On 3 of the 5 days for which the spectra were completed, signals <0.05 Hz (20 sec) were observed, indicating the presence of infragravity waves.

These spectral peaks almost universally demonstrated higher energy than other peaks found within the spectral signals; this trait was described by Ruessink (1998) as resulting from dissipation of short-period waves in the surf zone via breaking. Although there is no evidence that edge waves actually occurred, it seems possible

(a)

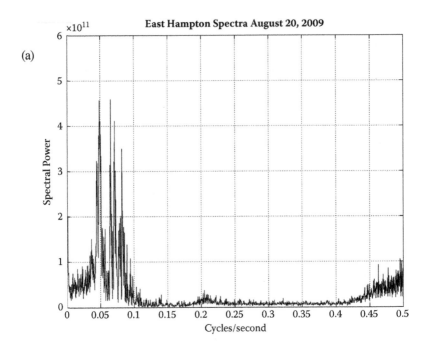

(b)

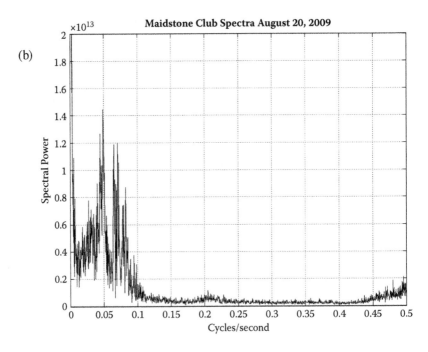

FIGURE 2.8 Spectral signals from (a) private residence and (b) Maidstone Club for seismic recordings along the Long Island south shore. Note the similar and dissimilar spectral peaks found in the records and clarified in Table 2.3.

TABLE 2.3

Periods of Spectral Peaks Observed at East Hampton Private Residence, Maidstone Club, and Average Period at NOAA Buoy 44017

Date (2009)	Private Residence	Maidstone Club	Average Period, Buoy 44017
July 17	17.04	34.48	
	4.65		4.5
	8.33		
	3.03		
August 20	4.878	4.878	4.8
	20.32	24.45	
	15.15	15.15	
	12.06		
	13.96	13.96	
August 22	50		
	7.69	7.69	6.54
	4.65	4.65	
	3.01	2.99	
October 2	5.13	4.69	4.02
	16.53	16.67	
		8	
October 6	40		
	3.68	3.57	3.85

that these frequencies may have generated resonant edge waves. The geometry of the nearshore between the headland created by the groins in the west and the offshore bar yields a calculated edge wave period between 70 and 100 sec, which is on par with the range of double the periods found in our spectra. Edge waves may have been driven by infragravity waves at these periods. More rip currents were not observed on the days when infragravity waves were detected in the spectra for East Hampton, but the existence of edge waves alone does not guarantee rip formation. These video records do not preclude the existence of rips during these periods of measured infragravity waves because the offshore-directed currents may have been out of the field of view.

CONCLUSIONS

Flash rip currents are characteristic of beaches on the south shore of Long Island. They are short lived (averaging 76 sec), and 2 to 13 rips occur per kilometer of shoreline per day. They are controlled by oceanographic processes rather than beach morphology. Edge waves may be responsible for these short-lived flash rip currents. While a direct relationship between the two phenomena remains unclear, the existence of spectral signals for longer period waves indicate the existence of infragravity waves.

REFERENCES

Aagaard, T.B., B. Greenwood, and J. Nielsen. 1997. Mean currents and sediment transport in a rip channel. *Marine Geology,* 140: 25–45.

Bleyer, W. 2009. Many South Shore ocean swimmers rescued from rip current. *Newsday,* July 31. http:/www.newsday.com/long-island/Suffolk/many-south-shore-ocean-swimmers-rescued-from-rip-current-1.1341834 (accessed August 3, 2009).

Bokuniewicz, H.J. 2004. Isolated groins at East Hampton, New York. *Journal of Coastal Research,* 33: 215–222.

Bowen, A.J. and D.L. Inman. 1969. Rip currents 2. Laboratory and field observations. *Journal of Geophysical Research,* 74: 5479–5491.

Brander, R.W. and A.D. Short. 2000. Morphodynamics of a large-scale rip current system at Muriwai Beach, New Zealand. *Marine Geology,* 165: 27–39.

Brewster, B.C., Ed. 1995. *The United States Lifesaving Association Manual of Open Water Lifesaving.* Upper Saddle River, NJ: Prentice Hall, pp. 43–45.

Bromirski, P.D. and F.K. Duennebier. 2002. The near-coastal microseism spectrum: spatial and temporal wave climate relationships. *Journal of Geophysical Research,* 107: 2166–2185.

Buonaiuto, F.S. 2003. Morphological evolution of Shinnecock Inlet, PhD dissertation, Stony Brook University, Stony Brook, NY.

Dalrymple, R.A. and C.J. Lozano. 1978. Wave-current interaction models for rip currents. *Journal of Geophysical Research,* 83: 6063–6071.

Darbyshire, J. and E.O. Okeke. 1969. A study of primary and secondary microseisms recorded in Anglesey. *Journal of Geophysical Research,* 17: 63–92.

Hammack, J., N. Scheffner, and H. Segur. 1991. A note on the generation and narrowness of periodic rip currents. *Journal of Geophysical Research,* 96: 4909–4914.

Hasselmann, K. 1963. A statistical analysis of the generation of microseisms. *Reviews of Geophysics,* 1: 177–209.

Holman, R.A., G. Symonds, E.B. Thornton et al. 2006. Rip spacing and persistence on an embayed beach. *Journal of Geophysical Research* 111: 1–17.

Kana, T.W. 1995. A mesoscale sediment budget for Long Island, New York. *Marine Geology,* 126: 87–110.

Komar, P.D. 1998. *Beach Processes and Sedimentation.* Upper Saddle River, NJ: Prentice Hall.

MacMahan, J.H., E.B. Thornton, and A.J.H.M. Reniers. 2006. Rip current review. *Coastal Engineering* 53: 191–208.

MacMahan, J.H., E.B. Thornton, T.P. Stanton et al. 2005. RIPEX: observations of a rip current system. *Marine Geology,* 218: 113–134.

Murray, A.B., M. LeBars, and C. Guilllon. 2003. Tests of a new hypothesis for non-bathymetrically driven rip currents. *Journal of Coastal Research,* 19: 269–277.

Oliver, J. and M. Ewing. 1957. Microseisms in the 11- to 18-second period range. *Bulletin of Seismological Society of America,* 47: 111–127.

Quartel, S. 2009. Temporal and spatial behavior of rip channels in a multiple-barred coastal system. *Earth Surface Processes and Landforms,* 34: 163–176.

Ruessink B.G. 1998. The temporal and spatial variability of infragravity energy in a barred nearshore zone. *Continental Shelf Research,* 18: 585–605.

Schubert, C. and H.J. Bokuniewicz. 1991. Infragravity wave motion in a tidal inlet. *Coastal Sediments.* American Society of Civil Engineers, 1434–1446.

Smith, J.A. and J.L. Largier. 1995. Observation of nearshore circulation: rip currents. *Journal of Geophysical Research,* 100: 967–975.

Turner, I.L., D. Whyte, and B.G. Ruessink et al. 2007. Observations of rip spacing, persistence and mobility at a long, straight coastline. *Marine Geology,* 236: 209–221.

West, M. and W. Meke. 2000. Fluid-induced changes in shear velocity from surface waves. Symposium on application of geophysics to engineering and environmental problems. Arlington, VA.

3 Rip Current Prediction at Ocean City, Maryland

Varjola Nelko and Robert A. Dalrymple

CONTENTS

INTRODUCTION

Several methods have been developed in recent years to predict rip currents at beaches in order to reduce risks to bathers (Lushine, 1991; Lascody, 1998). When the predicted risk is high, warnings are issued by the National Weather Service to increase public awareness and alert lifeguards. This chapter examines these models and presents a new rip current prediction methodology based on readily available environmental data on waves and tides for Ocean City, Maryland.

Ocean City is a nearly straight sandy beach on the Atlantic Ocean with a dynamic nearshore environment (Figure 3.1). Assessments of rip current threats (three times daily) were obtained from the Beach Patrol office in Ocean City and compared to the meteorological data obtained by nearby gages and buoys. The patterns in the data revealed by the Weka statistical software package show the conditions important for the generation of rip currents.

The driving force for rip currents is usually provided by alongshore variations in wave height arising for a variety of reasons including alongshore variable topography or spatially varying wave fields. The incoming incident waves carry mass, energy, and momentum into the surf zone where the waves dissipate by breaking. The radiation stresses of the waves defined by Longuet-Higgins and Stewart (1964) as "the excess flow of momentum due to the presence of waves" present a useful way to explain wave-induced phenomena in the surf zone such as the set-up of the mean water level, alongshore currents, and the interactions of currents with waves.

As waves approach the shore, they shoal and this results in an increase in wave height. On beaches with alongshore variations in bathymetry, the shoaling process

FIGURE 3.1   Rip current at Ocean City shown in still image taken from roof of Stowaway Grand Hotel as part of a Sea Grant-funded Erdman Video Systems installation of three video cameras and one still camera.

combined with other wave processes such as refraction and diffraction causes along-shore variations in wave heights. Wave breaking causes a decrease in wave momentum flux, producing compensating forces on the water column. An increase in the mean water surface provides a hydrostatic pressure gradient to counter the change in onshore wave momentum flux caused by breaking waves. As a result, the mean water surface displacement (set-up) increases (nearly) linearly in the shoreward direction and is proportional to the wave height (Bowen, 1969). In the case of wave-induced currents, the bottom frictional forces balance the change in the alongshore wave momentum flux. An alongshore variation in the set-up causes an additional driving force for alongshore currents. These currents transport water from areas of high set-up to areas of low set-up where rip currents are often located.

Relationships among the strengths of rip currents and high waves, tidal stages, and wave directions (Shepard et al., 1941; McKenzie, 1958; Sonu, 1972; Lushine, 1991; Dronen et al., 2002) are well known. Rip current magnitude has been found to increase with more shore-normal wave incidence, increasing wave height, and decreasing tide level. The offshore extent of rip currents is related to the height of incident waves (Shepard et al.,, 1941; Bowen, 1969); however, recent studies show that they can also be confined to the surf zone as part of a closed near-shore circulation cell (Reniers et al., 2007).

RIP CURRENT MECHANISMS

Many possible mechanisms may generate rip currents (MacMahan et al., 2006; Dalrymple et al., 2011). These currents are found on barred beaches (Figure 3.2)

FIGURE 3.2   Time exposure image depicting rips in Ocean City. The wave breaking patterns make it easy to distinguish the shallower regions that correspond to lighter intensity pixels and the channels that correspond to darker intensity pixels.

with rip channels (Sonu, 1966, 1972; Bowen, 1969; Noda, 1974; Dalrymple, 1978; Wright and Short, 1984; Lippman and Holman, 1990; Chen et al., 2004), cuspidal beaches (Hino, 1974), at beaches where coastal structures such as groins are present (Pattiaratchi et al., 2009), and at beaches with natural barriers such as islands and dredge holes (Mei and Liu, 1977; Pattiaratchi et al., 1987).

Other processes that can cause alongshore variation in wave height are wave interactions including the interaction of incident waves with edge waves (Bowen, 1969; Bowen and Inman, 1969; Guza and Davis, 1974; Sasaki et al., 1978; Symonds and Ranasinghe, 2000), synchronous trains of incident waves (Dalrymple, 1975), wave–current interactions (Le Blond and Tang, 1974; Mizuguchi and Horikawa, 1976; Dalrymple and Lozano, 1978; Falques et al., 2000; Yu and Slinn, 2003), wave groups (Dalrymple, 1975; Tang and Dalrymple, 1989; Fowler and Dalrymple, 1990; Haller et al., 1999; Kennedy and Dalrymple, 2001; Reniers et al., 2004), short-crested breaking (Peregrine, 1998; Johnson and Pattiaratchi, 2004), instability with waves, currents, and/or sediments (Hino, 1974; Dalrymple and Lozano, 1978; Dodd, 1996), and shear wave spawning (Ozkan-Haller and Kirby; Bowen and Holman, 1989; Reniers et al., 1997).

Since each of these mechanisms can generate rip currents, the local bathymetry and wave climate are critical in determining the types of rips that can occur at a given beach and which mechanisms may dominate. This implies that rip current prediction schemes may be site-specific.

DATA DESCRIPTION

Since July 26, 2008, trained lifeguards in Ocean City have classified rip threats three times daily. During the summer months of June to mid-September, the classification is as follows:

- Low Risk: Wind and/or wave conditions do not support the development of rip currents.
- Moderate Risk: Wind and/or wave conditions support the formation of rip currents in the surf zone; only some of these rips are dangerous; others are weaker in strength.
- High Risk: Wind and/or wave conditions support dangerous rip currents.

At high risk, the rips are very strong, and hundreds of rescues are performed daily. Every person swimming in a rip will be pulled offshore. Some days the rips are so strong that the beach is closed to swimming. In order to devise a numerical representation of rip risk data, the low risk data are given values of 1; moderate risk, 2; and high risk, 3. Meteorological data were obtained from the Engineer Research and Development Center (ERDC:http://chl.erdc.usace.army.mil/) of the U.S. Army Corps of Engineers. Some wind data came from the National Oceanic and Atmospheric Administration (NOAA; http://www.ndbc.noaa.gov/; Figure 3.3). The military gage MD002 is located offshore south of Ocean City and is the closest station to our area of interest. This station records wave height, wave period, wave direction, and tide data as defined on the ERDC website:

- Wave Height (H_{m0}): Spectrally derived wave height, in meters; equivalent to time domain-derived significant wave height in deep water. The height for Ocean City usually ranges from 0.2 to 1.4 m, but the waves can be as high as 3 m (see histogram in Figure 3.3).
- Wave Period (T_p): Peak spectral period, in seconds; inverse of the frequency of the peak (highest energy) of the one-dimensional power spectrum. The period of waves at the site usually ranges from 4 to 11 sec but at times can be as long as 18 sec.
- Wave Direction (D_p): Peak spectral direction, in degrees clockwise from true north; mean direction from which energy comes at the peak of the one-directional power spectrum.
- Depth: Nominal depth is the height of water above the pressure sensor; not referenced to a datum, but will serve as a proxy for tide level.

The meteorological data vary throughout the summer. Each month is character-ized by a different wave climate. For instance, May, the second half of August, and September have relatively higher waves that correspond to increasing rip threats. As Hurricane Bill approached the region late in August 2009, the lifeguards again issued a high rip threat that lasted for several days.

RIP CURRENT PREDICTION

An approach to predict rip currents at Ocean City, Maryland was developed by cor-relating the meteorological data against the lifeguard risk assessment. The rip threat assessment was preferred over the rip current rescue data because it represents the judgments of trained lifeguards and does not depend on the number of people on the beach or other factors that affect rescue numbers.

Previous forecasting techniques used the rip current rescue data as a measure of rip current activity, beginning with Lushine (1991) who developed Lushine Rip Current Scale (LURCS)—a forecasting technique to predict rip current danger in South Florida. The LURCS index utilized wind direction, wind speed, swell height, and time of low tide. Lascody (1998) modified LURCS for application to east central Florida beaches (ECFL LURCS). His model used wind direction, wind speed, swell period, swell height, and tide with modified weighting factors. Engle et al. (2002) analyzed the effects of wind, wave, and tide level on rip current occurrence for Daytona Beach and New Smyrna Beach in Florida and derived a scale with improved

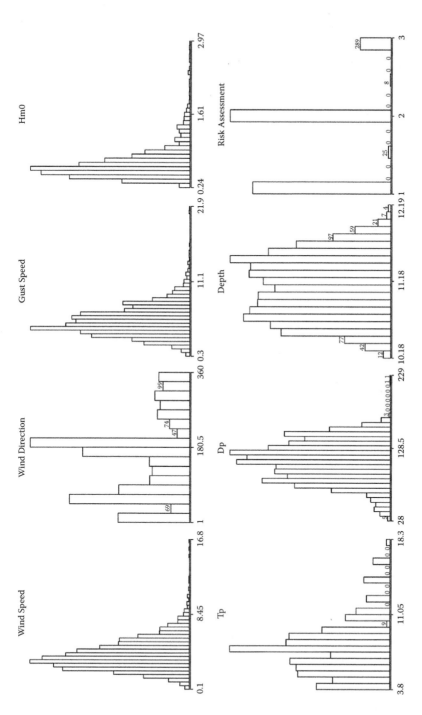

FIGURE 3.3 Probability distributions of wind speed, wind direction, gust speed, wave height, wave period, wave direction, and water depth along with rip assessments by lifeguards for all data (July 26 to September 21, 2008; May 23 to September 27, 2009; and May 29 to September 4, 2010).

accuracy (modified ECFL LURCS). Engle et al.'s scheme used wave height, wave period, wave direction, and tide as prediction parameters.

Schrader (2004) obtained wave data by deploying a directional gage at Daytona Beach and tested the modified ECFL LURCS checklist for the site. He found a relationship between rip current occurrence and pressure and frontal systems accompanied by strong winds and larger waves with long periods.

Another prediction scheme was developed by the National Weather Service (NWS) to predict rips for the mid-Atlantic beaches including Ocean City. Its scheme uses the same variables as LURCS and a similar table for risk calculation. The predictions of all methods are based on an easy-to-use table. For each variable in a scheme, a factor from the table that corresponds with the given magnitude of the variable is read. Summing these factors for all the variables yields a threat number. If the threat number exceeds a certain threshold, a warning is issued.

Because the wind and wave characteristics at beaches differ, it seems unlikely that a site-specific prediction scheme would exhibit the same predictive skill at another beach. This is further demonstrated by our attempt to predict rips at Ocean City using previously developed prediction schemes. The probability of detection (POD) ration and the false alarm ratio (FAR) are used to measure the accuracy of the calculation (Grenier et al., 1990; Lushine, 1991). The goal is to maximize the POD and minimize the FAR. These parameters are defined as:

$$POD = \frac{\text{number of rip warnings}}{\text{total number of rips}} \tag{3.1}$$

$$FAR = \frac{\text{number of false warnings}}{\text{total number of warnings}} \tag{3.2}$$

The total number of rips includes both moderate- and high-risk occurrences as evaluated by the lifeguards. The number of rip warnings is the total number calculated by the formula. For the correct prediction of all rip currents, POD = 100% and FAR = 0. Using the Ocean City data, the ECFL LURCS index calculated a POD of 12.45% and a FAR of 78.8%—a poor result. Engle et al.'s index (2002) or modified ECFL LURCS calculated POD as 5.7% and FAR as 15.9%. The LURCS, ECFL LURCS, and modified ECFL LURCS techniques developed for Florida beaches do not apply to Ocean City. The NWS index calculated a POD of 14.65% and a FAR of 6.25%. This does not predict many rips and results in a false low warning ratio.

DATA MINING AND CLASSIFICATION

Data mining is a branch of predictive analytics that deals with uncovering useful information from data and using it to predict future trends. Some of the earlier methods developed to discover patterns in the data include the Bayes theorem and regression analysis. Later methods include classification, clustering, and association rule learning. The different algorithms are more suitable to particular types of problems and determining which technique is suitable for a given problem is far from obvious.

A common data mining method is classification, which seems adequate for our purpose since it uses information from a variable in the data (in our case rip assessment by lifeguards) to classify the other variables in the data set that are believed to describe the chosen variable. The set of data used to uncover predictive relationships on future data is called a training set. A classifier maps the training data to a classification label. The algorithms common to the classification method include the decision tree (DT), neural network, k nearest neighbor (kNN), super vector machine (SVM), multilayer perceptron (MLP), naïve Bayes (NB), and RBF classifiers. Finding the most suitable classifier for a given set of data is no easy task and thus involves trial and error. Many authors (Van der Walt and Barnard, 2007) have studied the performances of classifiers using artificial data sets.

Many software systems have been developed to define data mining processes. Some open source systems include R Project, Weka, KNIME, and Rapid Miner. Weka (Waikato Environment for Knowledge Analysis), developed at the University of Waikato, New Zealand, was chosen. The Weka software is a collection of the state-of-the-art machine learning algorithms and data preprocessing tools. Weka-derived models can be used to generate predictions using new data.

Each row in our data set contains the date and hour, wind speed, wave height, wave period, wave direction, water level, and lifeguard risk assessment at the time of recording. The rows are called instances and the columns attributes. Four attributes and one outcome were used; wind speed was found to be unimportant. The attributes are further subdivided into (1) a training subset used to train the algorithms and (2) the test subset used to test the validity of the prediction scheme. The range of values of the attributes in the data set can be seen in Figure 3.3. Some attributes play a more significant role than others related to predicting instances in the data. The conditions for which rip current occurrence increases or decreases were sought. A simple way to represent the structural description of the information hidden in the data might be: if $H_{m0} > 1.4$ m, $T_p > 8$ sec, and depth <10.4 m, threat = 3. Satisfactory results were obtained from both the classification rule algorithm and the decision tree algorithm.

A common classifier in data mining is a univariate decision tree. A decision tree is constructed recursively. An attribute is selected for placement at the first node of the tree and a branch is built for each possible value of this attribute (Figure 3.4). These branches split the data into subsets. The algorithm proceeds recursively on each of the child nodes where again each node splits on only one attribute. The tree works from top to bottom, that is, from the first node to the leaves and the resulting tree best describes the data set. In the tree algorithm, the splits are evaluated using an information measure method that calculates the amount of information obtained by making a decision. The best tree chosen by regression is the one that gives the best information. These techniques are explained in more detail in Witten and Frank (2005).

In the category of decision tree classifiers, the M5 pruned model tree classifier performed very well, that is, the predicted threat correlated very well to the risk assessment by lifeguards. For this classifier, a complete decision tree is built, then pruning is performed by substituting a subtree with a leaf node; this is also called subtree replacement. Pruning decreases the size of the tree and makes the tree more readable. A complete tree is more accurate but is also more difficult to interpret. In

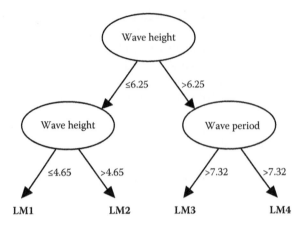

FIGURE 3.4 Simple pruned model tree that uses two variables (wave height and wave period) to describe lifeguard observations. A linear model is used to fit the data at the end of the branch.

the case of a pruned model tree, the predicted outcome for each replaced subtree is expressed as a linear sum of the significant attribute values with the appropriate weights chosen using standard regression. In our case, risk assessment by lifeguards for each end branch of the tree is expressed as a linear combination of all attributes (wave height, wave period, wave direction, and water depth) where the weights are the unknowns.

RESULTS

The prediction scheme resulting from the pruned model tree classifier begins with the wave height node:

$$\begin{cases} H_{m0} \leq 0.625 \text{ m} \Rightarrow \text{Threat} = 2.6345 * H_{m0} + 0.0214 * T_p - 0.1315 * \text{depth} + 1.4447 \\ \\ \\ H_{m0} > 0.625 \text{ m} \Rightarrow \text{Threat} = 0.5755 * H_{m0} + 0.0744 * T_p + 0.0031 * D_p + 0.7676 \end{cases}$$

$$(3.3)$$

$$(3.4)$$

if Threat $> 1.6 \Rightarrow$ issue a warning

The parameter that most clearly differentiates risk levels is wave height. We can conclude from these formulas that for smaller waves ($H_{m0} \leq 0.625$ m), wave height and water level play a significant role in rip threat. For higher waves, the respective weights applied to these variables change significantly and wave height and period are more important.

The threat or predicted rip risk is given by a linear model formula that results in a continuous function where the value for the training set ranges between 0.7 and 3.2

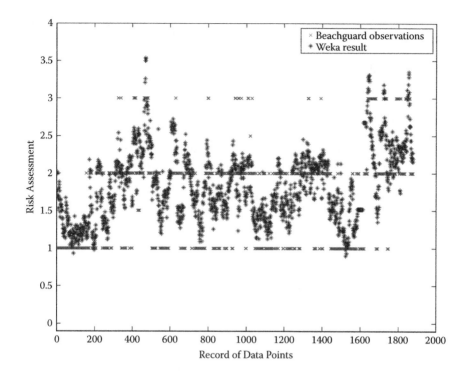

FIGURE 3.5 Weka formula results compared to rip threat assessments by lifeguards.

(Figure 3.5). The original rip assessment data by lifeguards is similar to a step function in which data are given one of three assigned values: 1 (low risk), 2 (moderate risk), or 3 (high risk). Here a value of the calculated threat was chosen for which the rips were considered dangerous and indicated a need to issue a rip warning. This is the threshold value. For a range of different values of this limit, the POD and FAR are calculated. The results are summarized in Table 3.1.

A threshold value of 1.6 is a reasonably good prediction with a corresponding POD of 84.8% and a FAR of 23.8%. A higher value could be chosen to decrease the number of false warnings; for instance, only a tenth of the warnings were false for a threshold of 2.0. However, a higher threshold also results in a lower number of warnings and this may create a false impression that a beach is safe when it is not. Above a threshold of 1.6, a rip warning should be issued and the prediction includes both moderate and high risks of rips. In the same way, only high-risk instances can be

TABLE 3.1

POD and FAR for Various Threshold Values

Threshold	1.2	1.3	1.4	1.5	1.6	1.7	1.8	1.9	2.0
POD	99.6	98.6	96.4	91.6	84.8	76.9	70.4	65.1	59.8
FAR	39.3	35.5	32.2	29.2	23.8	20.4	16.9	13.5	10.8

predicted; however, the formula could predict only a small percentage of these rips. Quantitatively, when a calculated threat is >2.7, a high rip threat is predicted. The calculation for high-risk rips results in a POD of 26% and a FAR of 16%.

To test the algorithm, the training and test subset approaches were used. The classifier was trained on randomly chosen sets of data to ensure that the formula was reproducible. Coefficients of the linear models for each training set are plotted in Figures 3.6 and 3.7. For each randomly chosen training set, the coefficients of the linear models are consistent. As for the testing part of the problem, the classifier was trained on summer 2008 and summer 2009 data (training subset). The resulting formula was then used to calculate the threat for summer 2010 data (test set), and the prediction resulted in a POD of 77% and a FAR of 25% (about 25% were predicted to be high-risk rips with no errors in prediction). As the size of the training set increases, the POD and FAR of the test set are expected to improve.

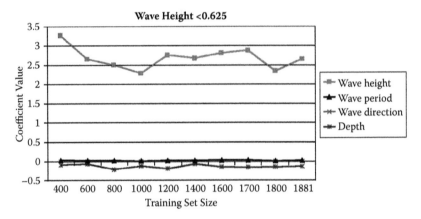

FIGURE 3.6 (*See color insert.*) Plot of coefficients of linear regression formula for first branch of pruned tree model (wave height <0.625) versus sample size of training set.

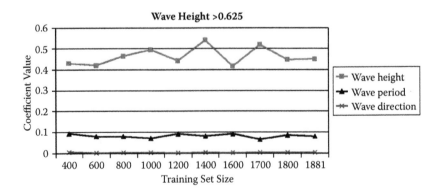

FIGURE 3.7 (*See color insert.*) Plot of coefficients of linear regression formula for second branch of pruned tree model (wave height >0.625) versus sample size of training set.

CONCLUSIONS

Wave, wind, and tide data from two consecutive summers at Ocean City, Maryland were studied. Previously prediction schemes for rip currents were not successful here, indicating that the scheme must be modified to suit different beach environments. Data mining software such as Weka can efficiently examine these types of trends. This information can be used by lifeguards to inform the public of threats from dangerous rip currents. The model will be refined further as additional data are obtained from lifeguards. A more generalized model based on data from other U.S. Atlantic beaches is in development as is an examination of the influence of beach stage as another variable (Nelko and Dalrymple, 2008).

ACKNOWLEDGMENTS

This work was initiated by a grant from the University of Maryland Sea Grant College Program (Prediction of Rip Currents: NA05OAR4171042). We would like to thank Captain Butch Arbin of the Ocean City Beach Patrol and the lifeguards for their help with the data. We also thank Professor Takeru Igusa for his help with data mining techniques.

REFERENCES

Arthur, R.S. 1962. A note on the dynamics of rip currents. *J. Geophys. Res.,* 67: 2777–2779.

Bowen, A.J. 1969. Rip currents 1. Theoretical investigations. *J. Geophys. Res.,* 74: 5467–5478.

Bowen, A.J. and Inman, D.L. 1969. Rip currents 2. Laboratory and field observations. *J. Geophys. Res.,* 74: 5479– 5490.

Bowen, AJ., Holman, RA. 1989. Shear instabilities of the mean longshore current, I. Theory. *J. Geophys. Res.,* 94: 18023–18030.

Brander, R.W. and Short, A.D. 2001. Flow kinematics of low-energy rip current systems. *J. Coast. Res.,* 17: 468– 481.

Chen, Q., Dalrymple, R.A., Kirby, J.T. et al. 1999. Boussinesq modeling of a rip current system. *J. Geophys. Res.,* 104: 20617–20637.

Dalrymple, R.A. 1975. A mechanism for rip current generation on an open coast. *J. Geophys. Res.,* 80: 3485–3487.

Dalrymple R.A. 1978. Rip currents and their causes. *Proc. 16th Coastal Eng. Conf.,* ASCE, Hamburg, 2: 1414–1427.

Dalrymple, R.A. and Lozano, C.J. 1978. Wave–current integration models for rip currents. *J. Geophys. Res.,* 83: 6063– 6071.

Dalrymple, R.A., MacMahan, J., Reniers, A.J.H.M. et al. 2011. Rip currents. *Annu. Rev. Fluid Mech.,* 43.

Dodd, N., Falques, A. 1996. A note on spatial modes in longshore current shear instabilities. *J. Geophys. Res.,* 101, C10, 22,715–22,726.

Dronen, N., Karunarathna, H., Fredsoe, J. et al. 2002. An experimental study of rip channel flow. *Coast. Eng.,* 45: 223– 238.

Engle, J., MacMahan, J., Thieke, R.J. et al. 2002. Formulation of a rip current predictive index using rescue data. Florida Shore and Beach Preservation Association National Conference.

Falques, A., Coco, G., and Huntley, D.A. 2000. A mechanism for the generation of wave-driven rhythmic patterns in the surf zone. *J. Geophys. Res.* 105: 24071–24087.

Fowler, R.E. 1991. *Wave Group Forced Nearshore Circulation: A Generation Mechanism for Migrating Rip Currents and Low Frequency Motion*. Final Research Report CACR-91-03. Center for Applied Coastal Research, University of Delaware, Newark.

Fowler, R.E. and Dalrymple, R.A. 1990. Wave group forced nearshore circulation. Proceedings of the 22nd International Conference on Coastal Engineering. Am. Soc. of Civ. Eng., Delft, pp. 729–742.

Grenier, L.A., Halmsted, J.T., and Leftwich, P. W. 1990. Severe Local Storm Warning Verification: 1989. NOAA Technical Memorandum NWS NSSFC-27. 22.

Guza, R.T. and Davis, R.E. 1974. Excitation of edge waves by wave incident on the beach. *J. Geophys. Res.,* 79: 1285–1291.

Guza, R.T. and Thornton, E.B. 1985. Observations of surf beat. *J. Geophys. Res.,* 90: 3161– 3171.

Haas, K.A. and Svendsen, I.A. 2002. Laboratory measurements of the vertical structure of rip currents. *J. Geophys. Res.,* 107.

Haller, M.C. and Dalrymple, R.A. 2001. Rip current instabilities. *J. Fluid Mech.* 433: 161–192.

Haller, M.C., Putrevu, U., Oltman-Shay, J., Dalrymple, R.A. 1999. Wave group forcing of low frequency surf zone motion. *Coast. Eng. J.,* 41: 121–136.

Haller, M.C., Dalrymple, R.A., and Svendsen, I.A. 1997. Rip channels and nearshore circulation. *Proc. Coast. Dyn.,* pp. 594–603.

Haller, M.C., Dalrymple, R.A., and Svendsen, I.A. 2002. Experimental study of nearshore dynamics on a barred beach with rip channels. *J. Geophys. Res.,* 107: 1 –21.

Hino, M. 1974. Theory on the formation of rip-current and cuspidal coast. *Proc. 14th Intl. Conf. on Coastal Engineering*. ASCE, Copenhagen, pp. 901– 919.

Holland, K.T., Holman, R.A., Lippmann, T.C. et al. 1997. Practical use of video imagery in nearshore oceanographic field studies. *J. Oceanic Eng.,* 22: 81–92.

Huntley, D.A. and Short, A.D. 1992. On the spacing between observed rip currents. *Coast. Eng.,* 17: 211– 225.

Johnson, D. and Pattiaratchi, C. 2004. Transient rip currents and nearshore circulation on a swell-dominated beach. *J. Geophys. Res.,* 109: C02026. doi:10.1029/2003JC001798.

Kennedy, A.B. and R.A. Dalrymple. 2001. Wave group forcing of rip currents. *Proc. Ocean Wave Measurement and Analysis*, ASCE, 1426–1435.

Lascody, R.L. 1998. East central Florida rip current program. *Natl. Weather Dig.,* 22: 25–30.

LeBlond, P.H. and Tang C.L. 1974. On Energy Coupling Between Waves and Rip Currents, *J. Geophys. Res.,* 79(6): 811–816.

Lippmann, T.C. and Holman, R.A. 1990. The spatial and temporal variability of sand bar morphology. *J. Geophys. Research,* 95: 11575–11590.

Longuet-Higgins, M.S. and Stewart, R.W. 1964. Radiation stress in water waves: a physical discussion with applications. *Deep Sea Res.,* 11: 529–563.

Lushine, J.B. 1991. A study of rip current drownings and weather related factors. *Natl. Weather Dig.,* 13–19.

MacMahan, J., Reniers, A.J.H.M., Thornton, E.B. et al. 2004. Infragravity rip current pulsations. *J. Geophys. Res.,* 109: C01033. doi:10.1029/2003JC002068.

MacMahan, J., Thornton, E.B., and Reniers, A.J.H.M. 2006. Rip current review. *J. Coast. Eng.,* 53: 191–208.

McKenzie, P. 1958. Rip current systems. *J. Geol.,* 66: 103–113.

Mei, C.C. and Liu, P.L.-F. 1977. Effects of topography on the circulation in and near the surf zone-linear theory. *J. Estuar. Coast. Mar. Sci.,* 5: 25–37.

Mizuguchi, M. and Horikawa, K. 1976. Physical aspects of wave-induced nearshore current systems, in *Proceedings of the 15th Coastal Engineering Conference*, pp. 607–625.

National Data Buoy Center. Historical Data, 2007–2009, NOAA/NDBC, available online.

Nelko, V. and Dalrymple, R.A. 2008. Rip currents: mechanisms and observations. In *Proc. 31st Int. Conf. Coast. Eng.,* Ed. J.M. Smith, pp. 888–900. Singapore: World Sci.

Noda, E.K. 1974. Wave induced nearshore circulation. *J. Geophys. Res.*, 79: 4097–4106.

Ozkan-Haller, T. and Kirby, J. 1999. Nonlinear evolution of shear instabilities of the long-shore current: a comparison of observation and computations. *J. Geophys. Res.*, 104: 25953–25984.

Pattiaratchi, C., Olsson, D., Hetzel, Y., and Lowe, R. 2009. Wave-driven circulation patterns in the lee of groynes. *Cont. Shelf Res.*, 29:1961–1974.

Pattiaratchi, C., James, A., and Collins, M. 1987. Island wakes and headland eddies: a comparision between remotely sensed data and laboratory experiments, *J. Geophys. Res.*, 92(C1), 783–794.

Peregrine, D.H. 1998. Surf zone currents. *Theor. Comput. Fluid Dyn.* 10: 295–309.

Reniers, A.J.H.M., Battjes, J.A., Falques, A., and Huntley, DA. 1997. A laboratory study on the shear instability of longshore currents longshore currents. *J. Geophys. Res.*, 102: 8597–8609.

Reniers, A.J.H.M, MacMahan J.H., Thornton E.B. et al. 2007. Modeling of very low frequency motions during RIPEX. *J. Geophys.Res.*, 112: C07013.

Reniers, A.J.H.M, Roelvink, J.A., and Thornton, E.B. 2004. Morphodynamic modeling of an embayed beach under wave group forcing. *J. Geophys. Res.*, 109: C01030. doi:10.1029/2002JC001586.

Sasaki, T. and Horikawa, K. 1978. Observation of nearshore current and edge waves. Coastal Engineering 1978. *Am. Soc. of Civ. Eng.*, Reston, Va, pp. 791–809.

Schrader, M. 2004. Evaluation of the modified ECFL LURCS rip current forecasting scale and conditions of selected rip current events in Florida. Master's thesis, University of Florida, Gainesville.

Shepard, F.P., Emery, K.O., and La Fond, E.C. 1941. Rip currents: a process of geological importance. *J. Geol.*, 49: 337–369.

Short, A.D. 1979. Three-dimensional beach stage model. *J. Geol.*, 553–571.

Sonu, C.J. 1972. Field observations of nearshore circulation and meandering currents. *J. Geophys. Res.*, 77: 3232–3247.

Symonds, G. and Ranasinghe, R. 2000. On the formation of rip currents on a plane beach. Proc. 27th Int. Conf. Coastal Eng. ASCE, Sydney, pp. 468–481.

Tang, E.C.S. and Dalrymple, R.A. 1989. Nearshore sediment transport. *Nearshore Circulation: Rip Currents and Wave Groups.* New York: Plenum Press.

Van der Walt, C.M. and Barnard, E. 2007. Data characteristics that determine classifier performance, *SAIEE Africa Res. J.*, 98: 87–93.

Witten, I.H. and Frank, E. 2005. *Data Mining: Practical Machine Learning Tools and Techniques*, 2nd ed. Morgan Kaufmann Series in Data Management, San Francisco. Available online: http://www.cs.waikato.ac.nz/~ml/weka/.

Wright, L.D. and Short, A.D. 1984. Morphodynamic variability of surf zones and beaches: a synthesis. *Marine Geol.*, 56: 93–118.

Yu, J. and Slinn, D.N. 2003. Effects of wave–current interaction on rip currents. *J. Geophys. Res.*, 108: 3088. doi:10.1029/2001JC001105.

4 Analysis of Rip Current Rescues at Kill Devil Hills, North Carolina

Gregory Dusek, Harvey Seim, Jeffrey Hanson, and David Elder

CONTENTS

INTRODUCTION

Rip currents constitute the number one cause for rescues and loss of life at United States beaches according to the U.S. Lifesaving Association. In 2007 alone, 40,810 of the 74,463 rescues reported at U.S. beaches were rip-related. Similarly, from a reported 109 drownings, 53 were caused by rip currents (www.usla.org). The status as the top beach safety hazard has garnered rip currents significant attention in the scientific research community. A plethora of research over the past decade has focused on a variety of topics: entire rip systems (MacMahan et al., 2005), rip current morphodynamics (Brander, 1999; Brander and Short, 2000; Calvete et al., 2005), rip current modeling (Garnier et al., 2008; Johnson and Pattiaratchi, 2006; Svendsen et al., 2000), surf zone bar behavior (van Enckevort and Ruessink, 2003; van Enckevort et al., 2004) and the relationship of rip currents to variability in local wave fields (Johnson and Pattiaratchi, 2004; MacMahan et al., 2004).

Despite the increase in rip current research, little investigation has concentrated on the large-scale alongshore (>1 km) and temporal (>1 month) variability of rip current activity. The likely reason for this research void is the difficulty in obtaining accurate observations of rip currents over large scales in time and space due to the complexity and cost of instrument deployment. As an alternative to instrument-collected observations, it is possible to use lifeguard rescue data as a relative indicator of hazardous rip current occurrence (Lushine, 1991; Lascody, 1998; Engle et al., 2002; Scott et al., 2007 and 2009).

This study utilizes a data set of 741 rip current rescues recorded at Kill Devil Hills, North Carolina. Each rescue is identified by both time and alongshore location and provides a unique opportunity to analyze large-scale variabilities in rip current activity. Concurrent with the rescue data, directional wave data, tidal height, and weather observations were collected at the nearby U.S. Army Corps of Engineers Field Research Facility in Duck. Additionally, cross-shore bathymetry profiles were collected along the length of Kill Devil Hills in 2008 and 2009. This chapter focuses on a statistical analysis of rip current rescues as correlated with tidal elevation and directional wave spectra to determine what factors most influence periods of increased hazardous rip current activity. Furthermore, analyses were performed to determine what factors influence variability in rip current activity, both temporally and alongshore.

RESEARCH UTILIZING RIP RESCUES

Multiple studies have used lifeguard rescue and drowning data as proxy information about rip current occurrence. The three studies discussed in detail below focus on predicting rip current occurrence through a statistical analysis of rip current rescues along with physical factors. Lushine (1991) was the first to attempt rip prediction when he analyzed the relationship of rip drownings in southeast Florida to a variety of meteorological and oceanographic data. He determined that rip current drownings were correlated with increasing wind speed, shore-normal wind direction, increasing wave height, and low tide. He used his results to create an empirical rip current forecasting or prediction index called LURCS (LUshine Rip Current

Scale), in which various inputs (wind speed and direction, wave height, tide) were assigned numerical values and added together to yield a rip current risk assessment. For example, 15-kt onshore winds, a 3-ft wave height, and low tide would indicate a category 5 risk or high likelihood for strong rip currents.

Lascody (1998) performed a similar analysis in east central Florida and with rip current lifeguard rescue instead of drowning data, thus providing a much larger data set. In addition to re-affirming that rip currents were correlated to wave height, low tide, wind speed, and wind direction, Lascody found that wave period was also a factor and that rip currents were more likely during long period swells (>12 sec). He formulated the ECFL (east central Florida) LURCS index that followed a method similar to Lushine's LURCS, with the addition of swell period as a factor. Four years later, Engle et al. (2002) performed additional analysis of lifeguard rescue data in east central Florida and made further changes to the ECFL LURCS index. They found that wind speed and direction were not important factors in determining rip current likelihood, but rather that wave field (peak period, peak direction, and height) and tide were the most accurate indicators of hazardous rip activity. Thus, a modified ECFL LURCS index utilizing these factors was created and successfully back tested. The modified index (or a slight variation) is the predominant rip current forecasting method used today by the National Weather Service's Weather Forecast Offices (WFOs).

The statistical relationship of tide, wave field, and rip current activity noted in the studies utilizing rip rescues has a physical basis identified in observational and numerical model studies. Previous observational studies (Brander and Short, 2000; MacMahan et al., 2005) determined that rip current intensity and activity are highest at low tide due to increased breaking over the surf zone bar at low water, leading to increased alongshore radiation stress gradients and greater current speeds as water is forced through rip channels from decreased water depth over the bar. Numerous observational studies find increasing rip current velocity with increasing wave height (Brander, 1999; Brander and Short, 2000; MacMahan et al., 2005), reflecting increased set-up and increased radiation stress gradients alongshore.

It is generally accepted that shore-normal wave incidence will lead to greater rip current activity (MacMahan et al., 2005) because highly oblique waves tend to drive stronger alongshore currents that can suppress rip current formation (Svendsen et al., 2000). However, relatively little observational research or numerical model research (Svendsen et al., 2000) demonstrates how rip activity varies with wave direction. Although some statistical analyses utilizing rip rescues have shown an increase in rip current activity with relatively long wave periods (Engle et al., 2002; Lascody, 1998; Scott et al., 2009), the relationship between rip activity and wave period is minimally addressed in the literature.

FIELD SITE

This study was performed at Kill Devil Hills, North Carolina—a 7.5-km stretch of beach located on the northern Outer Banks (Figure 4.1). The shoreline at the study site is generally straight and faces northeast with a shore-normal direction of 63 (±2) degrees true. The beaches of the northern Outer Banks are generally characterized by

FIGURE 4.1   Study location at Kill Devil Hills, North Carolina. Points show locations of 19 lifeguard chairs. Points with black dots mark locations where surf zone bathymetry was monitored in 2008 and 2009. The black cross-shore line indicates break between the 9 northern and 9 southern chairs.

a relatively steep foreshore (1:10), more gradual offshore (1:500), and shore-parallel bars (Schupp et al., 2006). The nearshore is often double-barred, with one bar in the surf zone (1 to 2-m depth) and an outer storm bar (~4.5-m depth) (www.frf.usace. army.mil). However, depending on the location alongshore and time of year, only one or no significant bars may be present. The mean annual significant wave height is 0.9 m (McNinch, 2004) and the wave climate is variable throughout the year.

The climate in the summer months, based on the observations used in this study, generally consists of a low energy swell of 0.4 to 0.6 m significant wave height (Hs) out of the southeast punctuated by storm events (1.0 to 3.0 m Hs), predominantly from the northeast. The tides are semi-diurnal and classified as microtidal as the mean range is 0.97 m (Birkemeier et al., 1985).

Kill Devil Hills (KDH) was chosen as the study site primarily due to the availability of nine summers of rescue data and the willingness of Kill Devil Hills Ocean Rescue to aid in the research. KDH Ocean Rescue occupies 19 lifeguard stands located 200 to 800 m apart along the beach. The stands are occupied from 10 am to 5:30 pm, seven days a week from late May until early September. Dating back to the summer of 2001, the rescue service maintains complete records of every lifeguard rescue made over the course of each summer. A record of each rescue indicates the type of rescue, the time, and location along the beach (in relation to the nearest lifeguard chair). Over the course of the nine summers of data collected, 741 rescues were classified as rip current related.

METHODS

DATA COLLECTION

This study assumed that for a rip current rescue to occur, a hazardous rip current was present at that particular location and time. In this case, a *hazardous* rip current is defined as having sufficient strength to cause a swimmer distress. It is important to note that little can be inferred from instances when no rescues were made. The fact that no rescues were made does not mean no rip currents were present. To the contrary, rip currents are likely on days of large surf conditions although they rarely involved rescues because most people do not go into the water or beaches may be closed to swimming. Similarly, stormy weather and cold water temperatures keep swimmers out of the water and these days are poorly represented in the rescue record.

Directional wave data, tidal heights, and bathymetry data were collected for correlation with the rescue data. Directional wave data were collected from a Waverider buoy maintained by the U.S. Army Corps of Engineers Field Research Facility (FRF). The buoy is located 15 km to the north of the study site at 17 m depth and was sampled hourly throughout the data period. During some time periods, back-up wave data were available from an FRF maintained Teledyne RD Instruments Acoustic Doppler Current Profiler (ADCP) located at 12 m depth at the northern extent of the study area. These data were only used as references and were not included in the statistical analysis due to the lack of a complete record. Data from the ADCP were used primarily to supplement the significant wave height time series from 2006; some Waverider data from that summer were missing. Tidal heights were observed from the pier at the FRF onshore of the Waverider buoy.

Bathymetry data were collected in the surf zone and in the outer nearshore. The surf zone is defined as the region from the beach seaward to just beyond the extent of breaking waves (~2 m depth) and the outer nearshore is the region beginning just outside the surf zone and extending 2 to 3 km offshore. Surf zone bathymetry data consist of profile lines at seven locations along KDH. Each profile line begins at a location slightly seaward of the dune line and transects at a shore-normal direction to about 2 m depth. The profile lines were re-occupied a number of times over the summer. Each line began at the same location and followed the same transect as closely as possible.

In 2008, data were collected along each profile line five times during the summer via a level and level rod at seven lifeguard chair locations. At each location, one transect was acquired. In 2009, bathymetry data were collected on four separate instances using a Trimble RTK GPS at the same seven locations utilized in 2008. While using the GPS, two profile lines 50 m apart were surveyed for each location. The vertical accuracy of the level and level rod is dependent on the distance from the level with the upper extent of the error at or near 10 cm. The vertical accuracy of the RTK GPS has a maximum error of 5 cm. Bathymetry data collected in the outer nearshore region consisted of a swath survey performed by the FRF in 2006. These data are considered to be reasonable estimates of the bathymetry over the study time period as the region's large-scale morphological features demonstrate relatively little short-term temporal variability (Schupp et al., 2006).

STATISTICAL ANALYSIS

Directional Wave Data

Data from the Waverider buoy are radio-telemetered to shore on a continuous basis. Spectral coefficients are computed onboard the buoy using the Fourier coefficient method (Longuet-Higgins et al., 1963) from contiguous 30-min records sampled at 1.28 Hz. Pawka's iterative maximum likelihood method (IMLM; 1983) is used to convert these observations to two-dimensional (2d) directional wave spectra. The significant wave height, peak period and peak wave direction is then calculated from the 2d wave spectra. Additional processing is performed on the wave spectra by the MATLAB® toolbox XWaves (www.WaveForceTechnologies.com). XWaves partitions 2d directional wave spectra into individual components through identification of spectral peaks and breaks by treating the spectra as inverted topographic domains and applying a watershed delineation transform. For a complete description, see Hanson and Phillips (2001), Hanson et al. (2009), and Tracy et al. (2006).

The classification of each component (wind sea—surface gravity waves forced by the local wind field or swell) is determined by the frequency and direction of each component relative to the local wind speed and direction. User options selected within XWaves determine how many spectral partitions are identified and how they are classified. For the Waverider data, a maximum of two partitions was allowed classified as either wind sea, dominant swell, or secondary swell and a minimum significant wave height of 0.2 m was required to identify a component. The significant wave height, peak period, mean wave direction, and directional spread were then calculated for each wave component.

Rescue Data Analysis

In prior studies utilizing rip current rescues to determine rip occurrence (Lushine, 1991; Lascody, 1998; Engle et al., 2002), histograms were used to compare overall conditions to conditions when rip rescues occurred. That same method of analysis is followed here with some adjustments and with further quantification of results.

For each data type, distributions of the entire data record and the rip rescue record were formulated. The entire data record consisted of the hourly observations from late May until early September from 2001 to 2009. The rip rescue record consisted of observations made during rip rescues, and if multiple rescues were made in an hour the events were counted accordingly (e.g., if three rescues occurred at 1 pm when the significant wave height was 1 m, the 1-m wave height was counted three times in the rip rescue distribution). It is important to note that the entire data record includes both daylight and evening hours, while the rescue record by its nature covers only daylight hours since no lifeguards are on the beach in the evening. Although wind sea can vary between daytime and evening hours due to the sea breeze–land breeze cycle, this variability was found to be slight when compiling the data for all nine summers and did not significantly alter interpretation of the results. Thus, to maintain data consistency and utilize as much of the data record as possible, both daylight and evening hours are included in the analysis. For visual analysis each distribution was plotted as a normalized histogram. If a particular physical property (e.g., peak period) had no impact on rip current activity, it would be expected that the histogram representing the rip rescue distribution would be similar to the entire data record distribution. Any significant deviation suggests that a property exerts some impact on rip occurrence. Additionally, where a deviation among histograms occurs suggests how that physical property impacts rip occurrence. Since rip current activity often depends on multiple wave spectral properties and wave direction, height, and period are often correlated, contour plots have also been created to visually represent and compare 2d distributions of data.

To test for significant differences between the entire data distribution and the rip rescue record distribution for each data type, the Kolmogorov-Smirnov (K-S) statistical test was used. The KS test, when applied to two empirical non-parametric distributions, can determine at a particular confidence level or p-value whether two distributions arise from the same underlying distribution (Conover, 1999). The KS test can be displayed graphically as confidence limits on two side-by-side cumulative distribution functions (CDFs) (Figure 4.2).

The sample distribution for each CDF can be said to represent the ensemble CDF within the confidence bounds with an x% certainty (99% as shown). If the confidence limits of the two distributions do not intersect everywhere, we can say within that level of certainty that the distributions are different. The test can also be shown numerically via the two-sample KS test that will result in a minimum p-value or maximum confidence level [$100 \times (1-p)$] for which the two distributions can be said to be different (Table 4.1). This method allows the variability of the distributions to be characterized with more detail and goes beyond a pass–fail for a set confidence level. For example, a KS test with a p-value of .03 and one with a p-value of 3×10^{-20} would both be said to be different at a p-value of .05 or a 95% confidence level, but a test resulting in a much smaller p-value can be said to be different with significantly more certainty (18 orders of magnitude). This information would be lost with a simple pass–fail measure of confidence.

RESULTS

INFLUENCE OF TIDAL ELEVATION AND WAVE FIELD ON RIP CURRENT ACTIVITY

Tidal Elevation

A comparison of the distributions from the KDH study shows evidence of increased rip activity at low tide levels (Figure 4.3). This result corresponds well to previous observational (Brander and Short, 2000; MacMahan et al., 2005) and statistical studies (Engle et al., 2002; Lascody, 1998; Lushine, 1991). For tidal elevations of

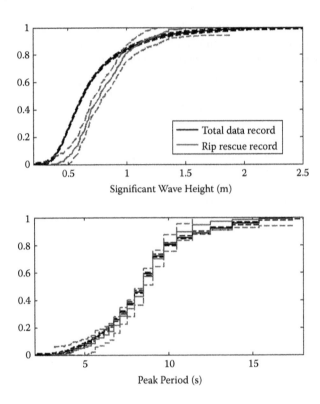

FIGURE 4.2 Cumulative distribution functions for significant wave height (top) and peak period (bottom) for the entire data record (black line) and rip rescue record (grey line). The 99% confidence intervals (dashed lines) are shown for each distribution.

0 m National Geodetic Vertical Datum (NGVD) and below, 58% of all rip rescues occurred, while only 36% of the entire data record was in this range (Table 4.2). Additionally, the p-value of the KS test is 3×10^{-33} essentially assuring that the two distributions are different (Table 4.1).

Bulk Measurements of Wave Field

Rip activity increases with increasing wave height and as wave direction is close to shore-normal (Engle et al., 2002; MacMahan et al., 2005; Svendsen et al., 2000). The histograms of significant wave height and peak direction from the KDH data set agree with previous research (Figure 4.4). Between significant wave heights of 0.6 m and 1.4 m, a substantial increase in rip current rescues was noted when compared to the entire data record (Table 4.2). This result suggests a strong relationship between wave height and hazardous rip current activity. At wave heights greater than 1.5 m, the slight decrease in the number of rescues can be attributed to adverse surf conditions and the unwillingness or inability of beachgoers to go into the water. The difference between the two distributions is almost certainly significant as the p-value is only 3×10^{-53} (Table 4.1).

TABLE 4.1
P-Values for Two-Sample Tests
of Rip Rescue Record and Entire
Data Record

Measurement	p-Value
Tidal elevation	3×10^{-33}
Entire Second Spectrum	
Significant wave height	3×10^{-53}
Peak period	0.021
Peak direction	5×10^{-57}
One Swell	
Significant wave height	9×10^{-45}
Peak period	0.003
Mean direction	6×10^{-83}
Directional spread	4×10^{-22}
Two Swells (Dominant)	
Significant wave height	2×10^{-10}
Peak period	0.052
Mean direction	3×10^{-17}
Directional spread	2×10^{-5}
Two Swells (Secondary)	
Significant wave height	8×10^{-10}
Peak period	9×10^{-6}
Mean direction	0.002
Mean direction (> 7s PP)	2×10^{-6}
Directional spread	2×10^{-6}
Wind Sea	
Significant wave height	0.030
Peak period	0.078
Mean direction	5×10^{-10}
Directional spread	0.099

The histogram of the peak direction shows that most of the wave energy arrives out of the southeast (>25 degrees south of shore-normal) during the summer months, while a majority of rescues (58%) occurred when the peak direction was within 25 degrees of shore-normal (Table 4.2). The p-value of 3×10^{-57} assures that the two distributions are different within a very high level of confidence. The histogram of the peak period shows less variability between the two distributions based on a maximum difference of only 4.9% at a period of 11 sec. Contrary to previous research, this suggests that period may not be an important factor when determining

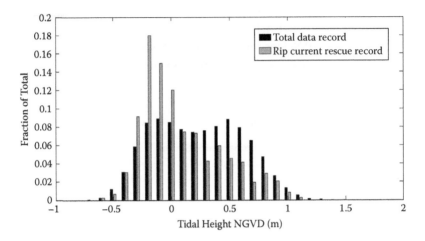

FIGURE 4.3 Distributions of tidal heights for the entire data record (black) and rip current rescue record (grey) represented as normalized histograms.

TABLE 4.2

Comparison of Occurrences of Various Factors in Rip Rescue Record and Entire Data Record

Measurement	Rip Record (%)	Entire Record (%)
Water level (<0 m NGVD)	58	36
Bulk Spectral Statistics		
Significant wave height (0.6 m < Hs < 1.4 m)	79	49
Peak direction within 25 degrees of shore-normal	58	30
Partitioned Spectra		
Only one or two swells present	77.2	64.9
Wind sea present	22.8	35.1
Event-Related Rescues		
72 hr following northerly event	40	19

rip current activity at KDH. Additionally, the KS test resulted in a p-value of .021, which provides relatively low certainty that the two distributions are different.

Partitioned Wave Spectrum

Once the wave spectral data are partitioned into individual components, evidence indicates that some components may play a larger role than others in hazardous rip current activities. Instances of either one or two swells and no measurable wind sea occurred 64.9% of the time for the entire data record, but 77.2% of the time for the rip current rescue record. Conversely, a wind sea is present 35.1% of the time in the entire record but only 22.8% of the time in the rip rescue record (Table 4.2).

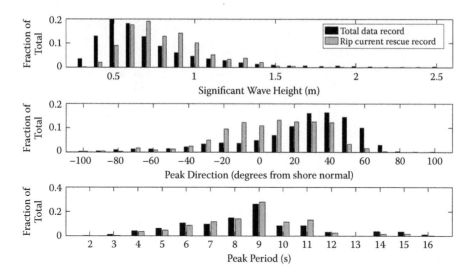

FIGURE 4.4 Distributions of significant wave height (top), peak direction (middle), and peak period (bottom) for the entire data record (black) and rip rescue record (grey). Distributions in each plot are shown as normalized histograms.

Although a lack of rescues does not necessarily indicate a lack of rip current activity, this disparity suggests that hazardous rip currents are more likely to occur during swell-dominated periods and not as likely when wind sea conditions are more significant. The p-values corresponding to the wind sea significant wave height and period are relatively large (Table 4.1)—providing additional evidence that for the wind sea component the rip rescue distributions are relatively similar to the distributions for the entire data record.

The p-value for the distributions of the wind sea mean direction is much smaller, almost assuring that the directional distributions are different. This may be because wind sea can have potentially large oblique angles of incidence that may be favorable for rip current development (MacMahan et al., 2006). Thus, the presence of wind sea at oblique angles may tend to suppress rip current activity due to the increased likelihood of a stronger alongshore current. This may further explain why fewer rescues occur when wind sea is present in the wave field.

Analysis of the swell components provides additional insight into potential wave field mechanisms for increased rip current activity. When only a single swell is present, hazardous rip currents are more likely to occur with higher significant wave heights and when the mean direction is closer to shore-normal (Figure 4.5). A wave vector time series from 2008 provides an example of increased rip current activity due to shore-normal single-swell conditions (Figure 4.6). In addition, the p-values for each statistic are extremely small (Table 4.1), affirming that the rip rescue distribution is different from the entire data set distribution at a confidence level well over 99%. This result correlates well with the analysis of the bulk spectral measurements. Similarities between the distributions of the single swell measurements and bulk spectral measurements were expected since the properties of the

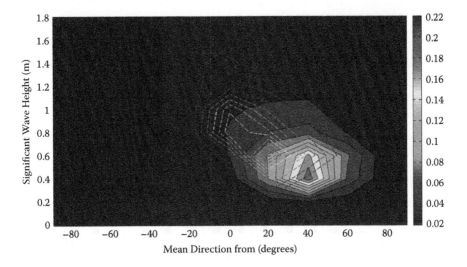

FIGURE 4.5 (*See color insert.*) Contour plots showing bivariate distribution of significant wave height and mean direction of swells when only this partition is present for the entire data record (solid) and rip rescue record (dashed). A mean direction from 0 represents shore-normal incidence where negative degrees appear north of shore-normal and positive degrees south of shore-normal. Contour values are fractions of totals for each distribution.

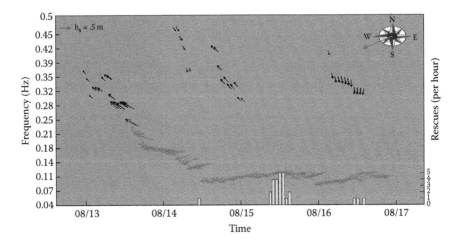

FIGURE 4.6 Wave vector plot over 4 days in 2008 showing a single-swell shore-normal wave field leading to increased rip current activity. Each vector represents frequency (vector origin), significant wave height (vector length), and wave direction (vector azimuth) of a spectral component for each hour. Light shaded vectors represent swell and dark vectors represent wind sea. A vector length equating to 0.5 m Hs is shown at upper left and the direction of shore-normal is shown by the arrow on the compass rose at upper right. The right y axis shows the number of rip rescues per hour, displayed as bars on the bottom of the plot.

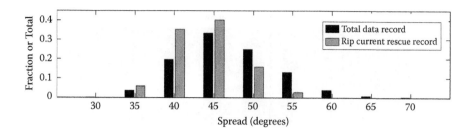

FIGURE 4.7 Normalized histograms representing distribution of directional spread for the entire data record (black) and the rip rescue record (grey) of the dominant swell.

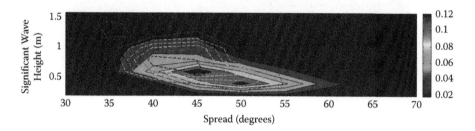

FIGURE 4.8 (*See color insert.*) Bivariate distribution of directional spread and significant wave height of dominant swell for the entire data record (solid) and rip rescue record (dashed). Contour values are the fractions of total for each distribution.

single swell component will be the same as the properties of the total spectrum in every instance of a single swell.

The directional spread of each swell component may also be a contributing factor to increased rip current activity. There is a noticeable increase in the relative number of rip rescues as the directional spread of the dominant swell decreases (Figure 4.7), and the very small p-value confirms that the distributions are different in this instance (Table 4.1). This increase in rescues is at least in part due to the relatively higher significant wave heights associated with smaller spread values (Figure 4.8). Despite this fact, the contour plots of the distributions suggest that the smaller spread plays at least a partial role in the increase in rescues. In the instances of multiple swells, rescues also increase with decreasing directional spread of each component, and this relationship appears to be significant in this case as well (Table 4.1).

For instances with two swells, the distribution of the rip rescue data looks very different from the entire data distribution (Figure 4.9). In the case of the dominant swell, the p-values for Hs and mean direction are extremely low (Table 4.1), indicating high levels of confidence that the rip rescue distributions are different from the entire data distributions. For the secondary swell, the p-value for the significant wave height was very small (8×10^{-10}), but the p-value for the mean direction, while small, is somewhat larger at 0.002. This is due to the secondary swell consisting of two different sources: long period swell from the southeast and short period swell, often from the northeast and with very small significant wave height, resulting

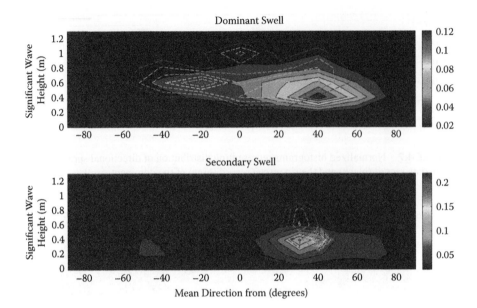

FIGURE 4.9 (*See color insert.*) Top: contour plot showing bivariate distribution of significant wave height and mean direction for the dominant swell component. Bottom: secondary swell component. In both cases, the entire data record (solid) and rip rescue record (dashed) are shown; for the mean direction, 0 represents shore-normal. Data obtained only when two swells and no measurable wind sea were present.

from local wind sea that is no longer wind forced. When the short period secondary swell (<7 sec) is removed, the p-value decreases by three orders of magnitude (Table 4.1).

The mean direction distributions of the rip rescue record for the instances of two swells are very different from the mean direction distributions of the rip rescue record when only one swell is present. For the two-swell case, the rip rescue distribution for the dominant swell is shifted toward more northerly directions and increased wave height, while the rip rescue distribution for the secondary swell is shifted toward more southerly directions and increased wave heights (Figure 4.9). The directional differences imply that rip activity increases when the dominant and secondary swells arrive at oblique angles with a large directional difference between them. A histogram of the swell direction difference confirms this and indicates a large increase in rip rescues when the difference of direction is between 60 and 100 degrees (Figure 4.10). This result suggests that a bi-directional spectrum representing crossing wave trains may be a mechanism for rip current generation. An example can be seen clearly from a 2004 wave vector plot (Figure 4.11). This possibility has been hypothesized (Dalrymple, 1978; Kennedy, 2005) and realized in numerical models (Johnson and Pattiaratchi, 2006) and laboratory studies (Fowler and Dalrymple, 1991). However, observational studies of the influence of multi-directional waves have been limited to instances of shore-normal wave incidence (Johnson and Pattiaratchi, 2004) and lack a full analysis of the directional spectra.

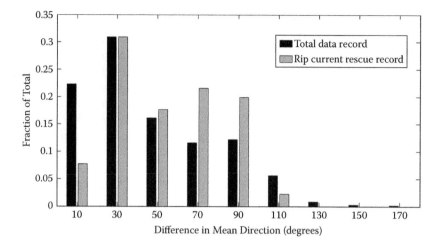

FIGURE 4.10   Normalized histograms representing distributions of swell mean direction difference when two swells were present. Direction difference is defined as the absolute value of the difference in the mean direction of the dominant and secondary swell components. Both the entire data record (black) and rip rescue record (grey) are shown.

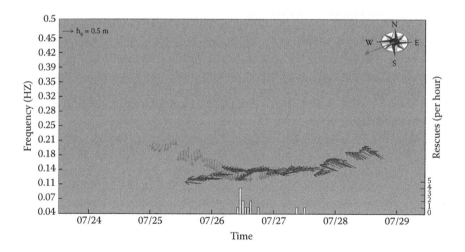

FIGURE 4.11   Wave vector plot over 4 days in 2004 showing a bimodal wave field leading to increased rip current activity. Each vector represents frequency (vector origin), significant wave height (vector length), and wave direction (vector azimuth) of a spectral component for each hour. Light shaded vectors represent swell 1 and dark vectors represent swell 2. A vector length equating to 0.5m Hs is shown at upper left and the direction of shore-normal is shown by the arrow on the compass rose at upper right. The right y axis shows the number of rip rescues per hour, displayed as bars on the bottom of the plot.

TEMPORAL VARIABILITY IN RIP RESCUES

The summer wave climate at KDH can be described as predominantly low energy swells (0.4 to 0.6 m Hs) out of the southeast, punctuated by storm events (1.0 to 3.0 m Hs), mostly from the northeast. The punctuated nature of the wave climate encourages a more detailed analysis of the effects of large wave events on rip current activity. For this analysis wave events are identified throughout the data record. The rescues following each event are compared with the event characteristics to determine the influence of large wave events on hazardous rip current occurrence.

Event Classification

Based on the summer wave climatology, events in which the significant wave height reached at least 1 m and lasted a minimum of 4 hr were identified. If a significant wave height dropped below 1 m for less than 4 hr before increasing over 1 m again, it was treated as the same event. A total of 115 events were identified over the 23,279 total hours of data collected over nine summers. This averages about 13 events per summer or roughly one event every 8.5 days. Events were classified as predominantly northerly or southerly relative to shore-normal. A total of 64 events exhibited average peak wave direction from north of shore-normal and 51 events were from the south. Events out of the north were typically front- or storm-related and thus began as wind sea events and transitioned into swell events following passage of the storm system. Events from the south were more often dominated by longer period swells. Event length ranged from the minimum 4 hr to a maximum of 129 hr. The average event length was 28.8 hours in duration.

The maximum significant wave heights of events varied from 1.02 m to a maximum of 3.45 m and averaged 1.56 m. The rescue period for each event consisted of the 72-hr window following the peak wave height of a particular event. In cases in which a rescue period overlapped with another event, the rescue period from the first event was cut short to prevent overlap with the rescue period of the second event.

Event-Related Rip Rescues

In many instances, a significant number of rescues were performed in the 72-hr periods following events. This was especially apparent in 2006 when each group of multiple rescue days followed shortly after a high-energy wave event (Figure 4.12). A total of 412 (or 56%) of the 741 rip rescues made over nine summers occurred during 72-hr windows following wave events. When a cross-correlation is made between the significant wave height and the hourly record of rip rescues over the entire data record, the maximum normalized value occurred when rescues were at a 21-hr lag from the significant wave height and the second and third highest values were noted at 45- and 68-hr lags, respectively; these lags essentially represent 1, 2, and 3 days following events.

It is important to note that the break in rescues between days does not represent a physical change in rip current activity but rather a decrease in bather load during evening hours (6 pm to 8 am). When considering the average peak wave direction of each event, 301 (or 73%) of the 412 rescues occurred in the rescue periods following events from north of shore-normal. The 72-hr rescue periods following northerly

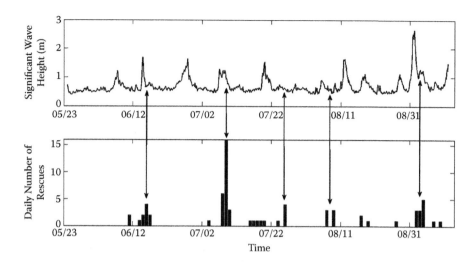

FIGURE 4.12 Top: hourly record of significant wave height for summer of 2006. Bottom: corresponding number of daily rip rescues. Arrows indicate high rescue days following large significant wave height events.

events account for a total of 4,344 hr of observations. Thus 40% of all rescues occur during only 19% of all observations (Table 4.2).

Increases in rip rescues following wave events may be wave field dependent, topographically controlled, or a combination of both factors since the typical characteristics of the wave field following an event are favorable to both rip current activity (MacMahan et al., 2005) and the development of an alongshore variable bar system (Calvete et al., 2005; Garnier et al., 2008; Lippmann and Holman, 1990). After an event, the wave field is typically characterized as having relatively high significant wave heights and close to shore-normal wave direction. Wave height, although steadily decreasing following the peak of the events, is still on average higher than during other time periods (0.92 m compared to 0.62 m). The dominant swell following wave events out of the north generally begins at a significant oblique angle from the north, trends toward shore-normal as wave energy decreases following the peak of the event, and eventually arrives from a slightly southerly direction (Figure 4.13). Thus, most of the time (63%) immediately following a northerly wave event, the dominant swell is within 25 degrees of shore-normal.

ALONGSHORE VARIABILITY

Variability in Number of Rip-Related Rescues

Data for the 741 rescues recorded from 2001 to 2009 includes the location of each rescue in relation to the nearest lifeguard chair and enables an analysis of the alongshore variability in the number of rip rescues at each chair location. From 2001 to 2008, a significant difference in the number of rescues made alongshore was noted (Figure 4.14). If the beach is divided into the northernmost nine chairs (~4 km) and the nine chairs to the south (~3 km) (Figure 4.1), a total

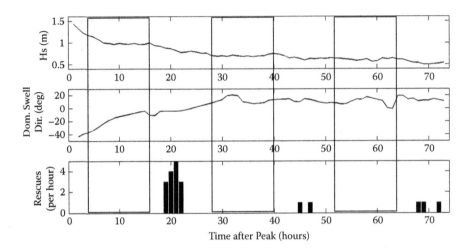

FIGURE 4.13 Plots show significant wave heights (top) and mean directions of the dominant swells relative to shore-normal (middle) and rip rescues per hour (bottom) for 72 hr after a large wave height event from the northeast. The x axis is the number of hours following the peak of the event and the rectangles represent the evening hours (1900 to 0600 EST) when rescues will not occur.

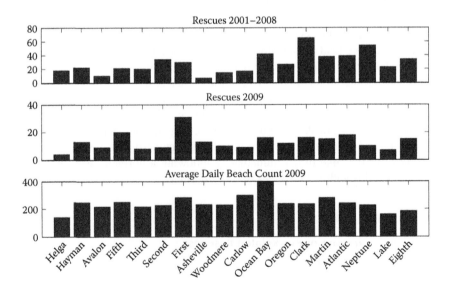

FIGURE 4.14 Top: total number of rip rescues made for each lifeguard chair (north to south) from 2001 to 2008. Middle: total number of rip rescues made for each chair in 2009. Bottom: average estimated daily beach count for each chair in 2009.

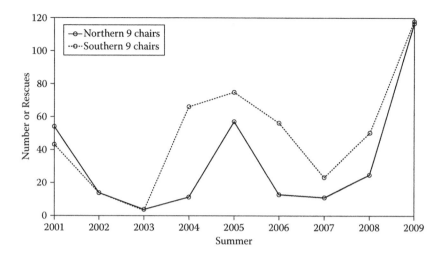

FIGURE 4.15   Number of rip rescues for each summer from 2001 to 2009. The dashed plot represents yearly total rip rescues for the nine northern lifeguard chairs and the solid plot shows the yearly total for the nine southern chairs.

of 177 rescues occurred in the northern half of KDH compared to 339 made in the southern half. In 2009, this disparity between north and south changed dramatically: 117 rescues in the northern half of KDH compared to 118 in the southern half.

Average daily beach counts recorded in 2009 show that beach attendance was relatively consistent alongshore, with the exception of Ocean Bay—the main beach access point in KDH. Although detailed beach count data are not available for other years, according to Ocean Rescue personnel, beach attendance is usually fairly uniform throughout KDH (personal communication). Since rescues are dependent on the number of people in the water, this demonstrates that beach attendance is not the primary reason for the variability in rescues alongshore. The number of rescues in 2009 and the distribution of these rescues alongshore appears to be relatively unique based on the annual variability of the northern and southern portions of KDH (Figure 4.15). From 2004 to 2008, the southern half of KDH consistently showed more rescues than the northern half, but that changed significantly in 2009. This result suggests that annual alongshore variability in the surf zone bathymetry may determine areas of increased hazardous rip current activity.

Role of Surf Zone Bathymetry

Surf zone bathymetry data are not available for the summers from 2001 to 2007 in the study area. However, cross-shore transects were performed at seven chair locations five and four times in the summers of 2008 and 2009, respectively. The generation of strong rip currents is closely tied to the surf zone morphology and more specifically to the extent of the surf zone bar system (Brander, 1999; Brander and Short, 2000; Haller et al., 2002; MacMahan et al., 2005). Thus, the expectation is that from 2004 to 2008 there would be more significant surf zone bar formation in

the southern half of KDH compared to the northern half and that in 2009 bar formation would be evident in most locations of KDH due to the large number of rescues
recorded at nearly every chair location.

The simple measure of bar presence in the profile lines recorded in 2008 and
2009 supports this expectation. It is important to note that the KDH region is often
double-barred, with an inner surf zone bar (1- to 2-m depth), and an outer storm bar
(~4.5 m-depth) (www.frf.usace.army.mil). This analysis is only of the inner surf zone
bar, as bathymetry data of the outer bar was not available and the outer bar is outside
of the surf zone region.

In 2008, the profiles recorded for the northern chairs (Hayman, Third, First,
and Asheville) rarely showed evidence of a surf zone bar in the measured region
(Table 4.3). Ocean Bay, which is counted among the nine southern chairs, also shows
no evidence of a surf zone bar. However, the profiles recorded at the two southernmost chairs (Clark and Neptune) show bars in four of the five dates when data were
collected. Comparing the profile lines recorded at First Street and Clark Street in
2008 demonstrates the significant difference in the surf zone bathymetry in the
northern and southern extents of KDH (Figure 4.16). While First Street exhibits
very linear profiles with no evidence of bar formation in the measured region, Clark
Street shows significant troughs and bars for four of the five profiling dates.

In 2009, the profiles recorded at every location along the beach nearly always
showed evidence of surf zone bars (Table 4.3). Comparing First Street and Clark
Street for the profiles performed in 2009 shows very different results from 2008
(Figure 4.17). The 2009 bathymetry at First Street is different from that in 2008 and
shows a significant trough and bar at every profile date. 2009 Clark Street data is
similar to from 2008 in that a significant bar system is evident at all four profile
dates. Consequently, in 2009, the bathymetry at First Street shows strong similarities
to the data collected at Clark Street. At both locations, the most significant trough
and bar were depicted on June 25, with a change to a more subtle surf zone bar by
July 15. This also suggests that, for the summer of 2009 (contrary to 2008), the
changes in the bar system alongshore at KDH are correlated in time.

TABLE 4.3
Fractions of Profiles Showing
Visible Bars for Chair Locations

Location	2008	2009
Hayman	0	0.75
Third	0.2	1
First	0	1
Asheville	0	1
Ocean Bay	0	1
Clark	0.8	1
Neptune	0.8	0.66

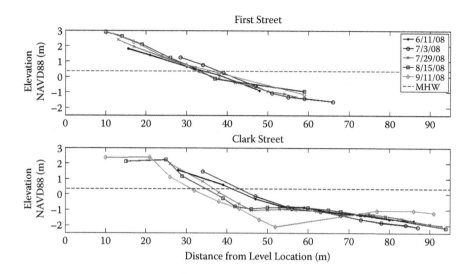

FIGURE 4.16   Cross-shore profiles of five instances in 2008 at First Street (top) and Clark Street (bottom) lifeguard chair locations. The x axis is the cross-shore distance from the starting GPS location, kept constant for each profile. The mean high water elevation of 0.36 m is shown (horizontal dashed) for each plot.

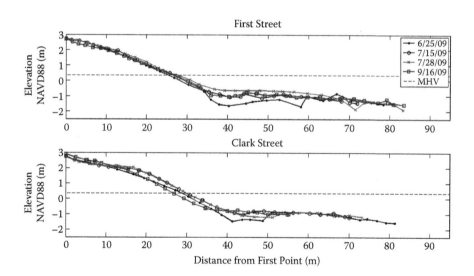

FIGURE 4.17   Cross-shore profiles of four instances in 2009 at First Street (top) and Clark Street (bottom) lifeguard chair locations. The x axis is the cross-shore distance from the starting GPS location, kept constant for each profile (same starting location used in 2008). Cross-shore distances are not equivalent for First and Clark Street so the mean high water elevation of 0.36 m is shown (horizontal dashed) for each plot.

It is evident that the surf zone bathymetry at KDH is fairly dynamic in both time and space, but also apparent is some alongshore persistence in the presence of the surf zone bar system. From the bathymetry data in 2008 and 2009, it can be inferred that the alongshore variability in the number of rip current rescues is related to the presence of a significant surf zone bar relatively near shore. The rescue record then suggests that while a strong bar system likely persisted in the southern portion of KDH in the summers from 2004 to 2008, there most likely was not consistent surf zone bar formation in the northern portion of KDH over the same time period. In 2009, significant bar formation was present along most of KDH for most of the summer. From 2001 to 2003, rescue numbers were similar in both segments, implying that bathymetric conditions were not as varied alongshore.

DISCUSSION

RIP-FAVORABLE WAVE CONDITIONS

An analysis of the partitioned wave data suggested two characteristic wave fields when hazardous rip currents are most favorable: (1) a single swell with relatively high significant wave height and shore-normal incidence; and (2) two distinct swells at highly opposing angles (>60 degrees) approaching at oblique incidence. The single swell case has been shown to be rip-favorable in previous studies (Engle et al., 2002; Svendsen et al., 2000) and describes the wave forcing often applied to numerical model (Calvete, et al., 2005; Svendsen et al., 2000) and laboratory studies (Haller et al., 2002). The bimodal case has received much less attention. Crossing wave trains have been shown to be potential mechanisms for rip currents in laboratory studies (Fowler and Dalrymple, 1991), but have never been documented observationally.

The importance of recognizing a bimodal wave field as a potential mechanism for hazardous rip currents is two-fold. First, rip currents of this nature are forced hydro-dynamically and thus do not rely on surf zone bathymetry (Johnson and Pattiaratchi, 2006). This fact may be especially significant in terms of rip current prediction, as rips forced from a bimodal wave field will not be constrained spatially and thus could occur anywhere along a shore. Second, during instances of two swells with highly opposing angles, the bulk statistics of the wave field will often represent a single wave direction at a highly oblique incidence. Thus, the present rip current forecast index, if it takes into account the wave direction, would predict low rip hazard conditions and be inaccurate in such cases because it relies on bulk spectral statistics like peak direction (Engle et al., 2002), and would not identify the secondary spectral peak.

SURF ZONE RESPONSES TO WAVE EVENTS

The temporal analysis of wave event-related rip rescues found that 40% of all rescues were made within 72 hr following wave events out of the northeast. The wave field following these events consisted of moderately high swells close to shore-normal that constitute wave conditions that are dynamically favorable for rip current activity (MacMahan et al., 2005). High energy and shore-normal swell conditions along with decreasing wave energy are also favorable for the development of an alongshore-variable surf zone bar system (Calvete et al., 2005; Garnier et al., 2008; Lippmann and Holman, 1990).

As rip currents are highly dependent on surf zone bathymetry (Brander, 1999; Haller et al., 2002; MacMahan et al., 2008), alongshore-variable surf zone bar systems will be associated morphodynamically with rip currents (Wright and Short, 1984). Thus, the increase in hazardous rip activity following these events is most likely due to wave conditions that are both favorable for rip current activity and for generating rip-favorable surf zone bathymetry. The occurrence of hazardous rip activity within 3 days following these events is also consistent with previous morphodynamic research.

It has been shown that immediately following a large wave event, a relatively alongshore-uniform bar develops on the outer boundary of the surf zone (van Enckevort and Ruessink, 2003; van Enckevort et al., 2004). As wave energy decreases, the bar moves toward shore, developing alongshore non-uniformities on roughly week-long time scales (van Enckevort et al., 2004). However, under moderate wave conditions, a partial reset is possible (van Enckevort et al., 2004), with such an event resulting in alongshore non-uniformities immediately following or within days following the moderate wave event (Garnier et al., 2008). This result corresponds well to the rip rescue record at KDH as wave events can typically be characterized as moderate (Hs of 1 to 2 m); most rescues occur 1 to 3 days following the event.

Additionally, since the surf zone bathymetry is closely tied to wave events, it is possible that wave events of similar magnitude and direction may result in a similar surf zone morphology following each instance. This hypothesis is especially significant to rip current forecasting, as often little information is available regarding surf zone bathymetry. If certain wave events force the surf zone bathymetry in such a manner that rip currents are more likely after these types of wave events, this factor could be included to improve the accuracy of rip current forecasts.

Nearshore Controls on Surf Zone Bathymetry

An analysis of the alongshore variability in rip rescues indicates that an increase in rip current activity is correlated to the presence of a surf zone bar, but why the surf zone bar system varies alongshore at KDH is uncertain. One possible explanation is the difference in the outer nearshore bathymetry (seaward of the surf zone) and underlying geology between the southern and northern portions of KDH. The nearshore of the northern Outer Banks is characterized by several regions of gravel outcrops and shore-oblique bars (McNinch, 2004), and these regions are typically correlated with paleo-river channels (Browder and McNinch, 2006). One such region begins north of KDH in Kitty Hawk and extends southward to near the location of First Street, covering over 3 km (Figure 4.18). The oblique bars extend at a northward angle from shore-normal and vary in size and scale. They can be found in depths as shallow as 2 m and reach >1 km from shore (McNinch, 2004). Additionally, the oblique bars and gravel outcrops are relatively stationary in location and time, showing essentially no variation following large wave events (Schupp et al., 2006).

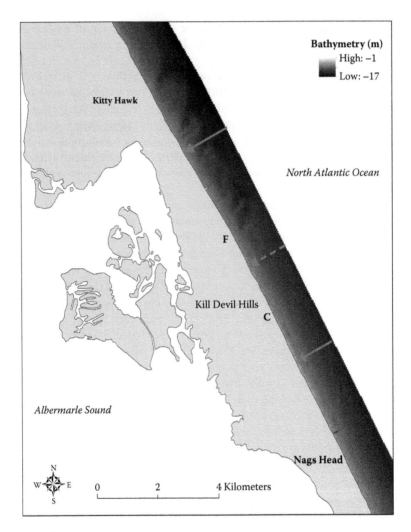

FIGURE 4.18 Data resulting from swath bathymetry survey performed by the U.S. Army Corps of Engineers Field Research Facility in 2006. Vertical scale is in meters NAVD88 and the northern and southern extents are shown (solid arrows). First Street (F) and Clark Street (C) are labeled. The northern nine chairs fall to the north of the dashed arrow; the southern nine chairs are south of the dashed arrow.

Although the bars do not extend into the surf zone, and thus do not directly influence surf zone processes, the morphological characteristics of this region are very different from southern KDH and may influence the behavior of the alongshore bar system. The northern region is characterized by a steeper and more variable cross-shore bathymetry gradient than exists in southern KDH (Figure 4.18). The northern region is also an area of relatively high rates of both short-term (1974–2002) and long-term (1933–1998) shore erosion and high rates of shoreline variability. Most of the southern portion has a relatively stable shoreline, experiences short-term accretion, and the entire region from First Street southward has net long-term accretion (Schupp et al., 2006). Furthermore, much of the northern region has a relatively thin and presumably active layer of sand compared to the southern region, which has a thicker and more uniform sand layer (Schupp et al., 2006).

These factors may all contribute to the variability in the alongshore (surf zone) bar system from northern to southern KDH. The steeper cross-shore slope in the northern portion presents a slightly more reflective beach state that results in an increase in wave energy in the surf zone and is less favorable to bar formation in general (Wright and Short, 1984). The relatively thin layer of sand and the erosion rates in northern KDH suggest a small sediment supply compared to southern KDH. A small sediment supply is also a characteristic of a more reflective beach state, and can also hinder surf zone bar formation. The reason for the sudden presence of a strong surf zone bar along essentially the entirety of KDH is more difficult to explain; however, the hypothesis posed by Schupp et al. (2006) that the upper sand layer in the northern portion is highly active supports the possibility of forming a significant bar system under optimal wave conditions.

CONCLUSIONS

The distribution of rip current rescues at Kill Devil Hills in both time and space suggests that rip current activity is dependent on the wave field and tide, previous wave conditions, and the surf zone bathymetry. In general these results conform well to previous research. The results of this study demonstrate that rip activity increases with increasing significant wave height, shore-normal wave incidence, and lower tidal elevation. These three factors have been shown repeatedly in previous research to impact rip activity and intensity (Brander, 1999; Brander and Short, 2000; Engle et al., 2002; MacMahan et al., 2005; Svendsen et al., 2000). It is also shown that rip currents are highly dependent on the surf zone bathymetry, which has been demonstrated in multiple publications as well (Brander, 1999; Brander and Short, 2000; Haller et al., 2002; MacMahan et al., 2005; MacMahan et al., 2008). However, the alongshore resolution and temporal extent of the rescue data combined with the availability of directional wave data and surf zone and nearshore bathymetry data have enabled a more detailed analysis of the contributions of these physical factors to rip current activity.

Analysis of the individual spectral components has shown that rip rescues are more likely to occur during swell conditions than when wind sea is present, but the relationship between rescues and the peak periods of the entire spectrum or the individual components is not very significant. When only a single swell is present in the spectrum, rip activity increases with increasing wave height, smaller directional spread, and wave incidence near shore-normal. When two swells are present, rip activity is most prevalent when the difference in the mean direction of each swell is between 60 and 100 degrees; this suggests that a bimodal wave field, causing crossing wave trains in the nearshore, may be an important mechanism for hazardous rip current occurrence.

Temporal data analysis demonstrated that rip currents are especially likely about a day after relatively large wave events from the northeast. Characteristics of the wave field following these events may be the primary reason for this increase in hazardous rip currents. The wave field within 72 hr following a northeasterly event generally consists of relatively large shore-normal swells. A wave field with these characteristics is not only dynamically favorable to rip current activity, but is also likely to generate alongshore-variable surf zone bathymetry, which itself increases the likelihood of hazardous rip activity.

Comparing the number of rip rescues alongshore at Kill Devil Hills suggests that rip currents are generally more likely in the southern half than in the northern half, but that this relationship can vary. From 2004 to 2008, a relatively high number of rip current rescues were made at KDH and for all those summers significantly more rescues were made in the southern half than in the northern half. In 2009, the number of rescues was well above average along the entirety of the shore. An analysis of the surf zone profiles recorded in 2008 show that the southern half of KDH had surf zone bar formation throughout the summer, consistent with the high number of rip rescues, while the northern half did not. Since rip rescue records from 2008 are consistent with those from the previous 4 years, this is presumed to be the normal mode of surf zone bathymetry alongshore. However, this mode is subject to variability for a particular summer and in 2009 the bathymetry varied dramatically. Additionally, the presence or lack of surf zone bar morphology at a particular location alongshore appears consistent over the course of one summer.

Although the results of this study provide valuable insight into how hazardous rip current activity is influenced, it is important to note the limitations of using rescue data as the primary rip current observational resource. As mentioned previously, not having a rip current rescue at a particular location and time indicates very little regarding whether a hazardous rip current existed at that location and time. Rip rescues are closely tied to bather load, and if bather load is low due to bad weather, cold water temperatures, or beach closures, few or no rescues will be made even if hazardous rip currents are present. To address this concern, lifeguard observations of rip current activity and intensity were made in 2008 and 2009, and these data will be included in a future study to verify the current results.

ACKNOWLEDGMENTS

Dusek was supported by NOAA IOOS Grant NA07NOS4730212. This material is based in part upon work supported by the U.S. Department of Homeland Security under Award 2008-ST-061-ND 0001. The views and conclusions contained in this document are those of the authors and should not be interpreted as necessarily representing the official policies, expressed or implied, of the U.S. Department of Homeland Security.

Waverider data and swath bathymetry data were provided by the Field Research Facility, Field Data Collections and Analysis Branch, U.S. Army Corps of Engineers, Duck, North Carolina. Field work performed at Kill Devil Hills was supported by U.S. Army Corps of Engineers Grant W912BU-9-P-0236. Additionally, the field work performed for this research could not have been possible without support from the U.S. Army Corps of Engineers Field Research Facility, Kill Devil Hills Ocean Rescue, and the University of North Carolina Coastal Studies Institute.

REFERENCES

Birkemeier, W.A., Miller, H.C., Wilhelm, S.D. et al. 1985. *A User's Guide to the Coastal Engineering Research Centers (CERC's) Field Research Facility.* Coastal Engineering Research Center, U.S. Army Corps of Engineers, Vicksburg, MS.

Brander, R.W. 1999. Field observations on the morphodynamic evolution of a low-energy rip current system. *Marine Geol.,* 157: 199–217.

Brander, R.W. and Short, A.D. 2000. Morphodynamics of a large-scale rip current system at Muriwai Beach, New Zealand. *Marine Geol.,* 165: 27–39.

Browder, A.G. and McNinch, J.E. 2006. Linking framework geology and nearshore morphology: correlation of paleo-channels with shore-oblique sandbars and gravel outcrops. *Marine Geol.,* 231: 141–162.

Calvete, D., Dodd, N., Falques, A. et al. 2005. Morphological development of rip channel systems: normal and near-normal wave incidence. *J. Geophys. Res.,* 110: C10006.

Conover, W.J. 1999. *Practical Nonparametric Statistics.* John Wiley & Sons, New York, pp. 428–473.

Dalrymple, R.A. 1978. Rip currents and their causes. *Proc. 16th Conf. on Coastal Eng., Vol. II,* 1414–1427.

Engle, J., MacMahan, J., Thieke, R.J. et al. 2002. Formulation of a rip current predictive index using rescue data. Florida Shore and Beach Preservation Association National Conference.

Fowler, R.E. and Dalrymple, R.A. 1991. Wave group forced nearshore circulation. *Proc. 22nd. Intl. Conf. on Coastal Eng.,* 729–742.

Garnier, R., Calvete, D., Falques., A. et al. 2008. Modelling the formation and the long-term behavior of rip channel systems from the deformation of a longshore bar. *J. Geophys. Res.,* 113: C07053.

Haller, M.C., Dalrymple, R.A., and Svendsen, I.A 2002. Experimental study of nearshore dynamics on a barred beach with rip channels. *J. Geophys. Res.,* 107: 1–21.

Hanson, J.L. and Phillips, O.M. 2001. Automated analysis of ocean surface directional wave spectra. *J. Atmos. Oceanic Technol.,* 18: 277–293.

Hanson, J.L., Tracy, B.A., Tolman, H.L. et al. 2009. Pacific hindcast performance of three numerical wave models. *J. Atmos. Oceanic Technol.,* 26: 1614–1633.

Johnson, D. and Pattiaratchi, C. 2004. Transient rip currents and nearshore circulation on a swell-dominated beach. *J. Geophys. Res.,* 109: C02026.

Johnson, D. and Pattiaratchi, C. 2006. Boussinesq modeling of transient rip currents. *Coastal Eng.* 53: 419–439.

Kennedy, A.B. 2005. Fluctuating circulation forced by unsteady multidirectional breaking waves. *J. Fluid Mech.,* 538: 189–198.

Lascody, R.L. 1998. East central Florida rip current program. *Nat. Weather Dig.* 22: 25–30.

Lippmann, T.C. and Holman, R.A. 1990. The spatial and temporal variability of sand bar morphology. *J. Geophys. Res.,* 95: 11575–11590.

Longuet-Higgins, M.S., Cartwright, D.E., and Smith, N.D. 1963. Observations of the directional spectrum of sea waves using the motions of a floating buoy. *Ocean Wave Spectra.* Prentice-Hall, Englewood Cliffs, NJ, pp. 111–136.

Lushine, J.B. 1991. A study of rip current drownings and related weather factors. *Nat. Weather Dig.,* 16: 13–19.

MacMahan, J.H., Reniers, A.J.H.M., Thornton, E.B. et al. 2004. Infragravity rip current pulsations. *J. Geophys. Res.,* 109: C01033.

MacMahan, J.H., Thornton, E.B., Stanton, T.P. et al. 2005. RIPEX: observations of a rip current system. *Marine Geol.,* 218: 113–134.

MacMahan, J.H., Thornton, E.B., and Reniers, A.J.H.M. 2006. Rip current review. *Coastal Eng.,* 53: 191–208.

MacMahan, J.H., Thornton, E.B., Reniers, A.J.H.M. et al. 2008. Low-energy rip currents associated with small bathymetric variations. *Marine Geol.*, 255: 156–164.

McNinch, J.E. 2004. Geologic control in the nearshore: shore-oblique sandbars and shoreline erosional hotspots, Mid-Atlantic Bight, USA. *Marine Geol.*, 211: 121–141.

Pawka, S.S. 1983. Island shadows in wave directional spectra. *J. Geophys. Res.*, Vol. 88, 2579–2591.

Schupp, C.A., McNinch, J.E., and List, J.H. 2006. Nearshore shore-oblique bars, gravel outcrops, and their correlation to shoreline change. *Marine Geol.*, 233: 63–79.

Scott, T., Russell, P., Masselink, G. et al. 2007. Beach rescue statistics and their relation to nearshore morphology and hazards: a case study for southwest England. *Intl. Coastal Symp. J. Coastal Res., Special Issue,* 50: 1–6.

Scott, T., Russell, P., Masselink, G. et al. 2009. Rip current variability and hazard along a macro-tidal coast. *Intl. Coastal Symp. J. Coastal Res., Special Issue,* 56: 895–899.

Svendsen, I.A., Haas, K.A., and Zhao, Q. 2000. Analysis of rip current systems. *Coastal Eng. Proc. 27th Intl. Conf.,* ASCE, New York, 1127–1140.

Tracy, F.T., Tracy, B.A., and Hanson, J.L. 2006. Sorting out waves with a fast sort algorithm. *ERDC MSRC Resource.* U.S. Army Engineer Research and Development Center, Vicksburg, MS.

van Enckevort, I.M.J. and Ruessink, B.G. 2003. Video observations of nearshore bar behaviour. Part 1: alongshore uniform variability. *Cont. Shelf Res.,* 23: 501—512.

van Enckevort, I.M.J. and Ruessink, B.G. 2003. Video observations of nearshore bar behaviour. Part 2: alongshore non-uniform variability. *Cont. Shelf Res.,* 23: 513–532.

van Enckevort, I.M.J., Ruessink, B.G., Coco, G. et al. 2004. Observations of nearshore crescentic sandbars. *J. Geophys. Res.,* 109: C06028.

Wright, L.D. and Short, A.D. 1984. Morphodynamic variability of surf zones and beaches: a synthesis. *Marine Geol.*, 56: 93–118.

5 Methodology for Prediction of Rip Currents Using a Three-Dimensional Numerical, Coupled, Wave Current Model

George Voulgaris, Nirnimesh Kumar, and John C. Warner

CONTENTS

INTRODUCTION

Rip currents constitute one of the most common hazards in the nearshore that threaten the lives of the unaware public that makes recreational use of the coastal zone. Society responds to this danger through a number of measures that include: (a) the deployment of trained lifeguards; (b) public education related to the hidden hazards of the nearshore; and (c) establishment of warning systems.

The U.S. National Oceanic and Atmospheric Administration (NOAA) has teamed with the U.S. Lifesaving Association (USLA) to undertake outreach activities aimed at informing the public about the hazards of rip currents. In addition, the National Weather Service (NWS) established the Coastal Weather Forecasts (CWFs) and the Surf Zone Forecasts (SRFs) produced by local Weather Forecast Offices (WFOs). CWFs are produced by all WFOs and are used by emergency managers, the media, and the general public for planning purposes to support and promote safe transportation across the coastal waters.

SRFs are not mandatory and their issuance is determined by regional policies and customer and partner needs (NOAA, 2004). NOAA (2004) states that SRFs were established in recognition of the fact that such a forecast "…provides valuable and life-saving information, pertaining to hazards in the surf zone, to the beach front community, including the general public, and providers of beachfront safety services, such as lifeguards." Even if a WFO does not issue an SRF, it may elect to provide rip current information within a CWF using a three-tiered text qualifier: low, moderate, and high risks (NOAA, 2004).

Currently, rip current forecasting is based on predictive indices that in turn are based on statistical correlations between rip current-related rescues and information about wave and wind conditions. Lushine (1991) developed the first index for South Florida, known as the Lushine Rip Current Scale or LURCS. This index is an empirical forecasting technique that utilizes wind direction and speed, swell height, and timing of low tide to predict rip current dangers. It was later modified by Lascody (1998) for use in east central Florida (ECFLS LURCS).

The modification included the addition of swell period and a slight modification of the tidal factor. Engle et al. (2002) reviewed lifeguard rescue logs from Daytona Beach, Florida, and found that the frequency of rip current rescues increased during shore-normal wave incidence and during mid-low tidal stages. They removed the wind speed and direction information from the rip current risk prediction index and added tidal stage and wave direction. Their numerical prediction index (modified ECFLS LURCS) assigns different weights to different wave parameters. The total values vary from 0 to 12 indicating low (value 0) to very high (value 12) rip current risk.

Engle et al. (2002) noted that correlations between rescues and environmental conditions are not reliable means of determining rip dangers. Beach attendance reduces with increased wave energy and bad weather. Thus such statistical samples are biased if not normalized based on the total population visiting the beach. In addition, fatalities or rescues caused by factors other than rip currents (e.g., swimming competence level, alcohol consumption, medical condition, alongshore current strength, etc.) are not included in this approach. As a result, all beach rescues are attributed to rip currents, thus biasing the statistics. Also, the rip prediction indices do not include data on beach morphology (alongshore bars, ridge and runnel morphology, etc.) although scientific evidence suggests that rip currents are related to the presence of longshore bars (Haller et al., 2002; Aagaard et al., 1997; Wright and Short, 1984; Sonu, 1972).

It can be argued that beach morphology is included indirectly, since the statistical correlations are based on data for a particular beach with a particular morphology.

This argument, although valid, reinforces the fact that the risk index developed strongly depends on local characteristics and its use in other locations with potentially different morphologies may lead to inaccurate forecasts. Despite its empirical nature, the above-described rip current hazard index relies on information on wind and wave conditions—two important physical forces that contribute to rip current generation.

The NWS currently provides weather (wind) forecasting for coastal areas. Wave forecasts provided by NOAA's National Center for Environmental Prediction (NCEP) are limited in resolution (1.25 degrees × 1 degree and 0.25 degree × 0.25 degree for the global and western north Atlantic, respectively) and do not resolve nearshore bathymetry or coastline variabilities that may modify nearshore wave dynamics significantly. Increased resolution wave forecasting models have been implemented locally by a limited number of NWS WFOs including Eureka, California; Wakefield, Virginia; and Newport, Morehead City, and Wilmington, North Carolina (Devaliere et al., 2009; Willis et al., 2010) that improve the predictive capability within the middle and inner shelves. However, no forecast covers areas inside the surf zone.

Allard et al. (2008) presented a methodology for predicting longshore currents using commercial software (DELFT-3D) in a depth-averaged mode, but it suffered from lack of information on bathymetry, and it did not address the issue of rip currents directly. This contribution attempts to fill this gap by incorporating latest developments of nearshore hydrodynamics and numerical modeling, using public domain 3-D code, toward the development of a quantitative rip current prediction tool.

BACKGROUND ON RIP CURRENT DEVELOPMENT

As waves travel in the nearshore, they are modified through refraction, diffraction, dissipation, and shoaling processes, but can also gain energy through local wind forcing (Kinsman, 1965; Lavrenov, 2003). In very shallow water depths, wave heights tend to be controlled by local water depth through the process of wave breaking. The excess of momentum due to the presence of waves (radiation stress, Longuet-Higgins and Stewart, 1964) is generally conserved until the breaking point at which the energy is expended and wave-induced currents are also generated. In the nearshore, two wave-induced current systems are described (Komar 1986): (1) alongshore currents generated by obliquely approaching waves; and (2) a cell circulation system (Shepard and Inman, 1950) that occurs when the incident angle of the waves is small, and usually is responsible for the generation of rip currents.

The latter circulation system consists of (a) onshore transport of water, (b) longshore currents that are confined within the surf zone and carry water toward the rip (feeder currents); (c) the rip neck, a fast-flowing current that extends from the point of confluence of the two opposing feeder currents and transports water through the surf zone; and (d) the rip head, which is a region of decreasing velocity seaward of the surf zone.

Rip currents are integral parts of the cell circulation system. They are jet-like and directed seaward with speeds that can reach up to 2 ms⁻¹ under extreme conditions (Short, 1985) but with typical values of 0.5 to 1 ms⁻¹. They are usually associated with intermediate (Wright and Short, 1984) type beaches with bar and rip morphologies (Sonu, 1972), although recent theories (Murray et al., 2003) suggest that rip currents can also occur on alongshore-uniform beaches as self-organized features (flash rips).

The ability to model nearshore circulation and rip currents improved through the introduction of the concept of radiation stress by Longuet-Higgins and Stewart (1964). According to this theory, shoreward of the wave breaking zone, mean sea level increases (wave set-up), generating a pressure gradient that balances the gradient due to radiation stress. Assuming laterally uniform nearshore morphology, if the waves approach the coastline at an angle, then the alongshore component of the excess of momentum drives alongshore currents that flow parallel to the coastline (Longuet-Higgins, 1970; Thornton, 1970). Their strength depends greatly on their angle of approach and wave breaking height. Under large angles of approach, longshore current velocities can exceed 1 ms^{-1} and can be dangerous for swimmers, especially children.

These currents have been studied extensively since they are responsible for transporting sediment in the alongshore, causing coastal erosion. It is generally agreed that breaking waves stir sediment (Voulgaris and Collins, 2004) and the longshore current transports it along the coast (Voulgaris et al., 1996; 1998). An alongshore gradient of this longshore sediment transport is responsible for the development of erosional and/or accretional areas along a coastline.

Overall, higher waves will create higher set-ups (Bowen, 1969) and a lateral variation in the wave height will create a lateral variability in the wave set-up, developing an alongshore pressure gradient (Bowen, 1969; Dalrymple, 1978) that can drive longshore (feeder) currents from areas of high waves toward areas of low wave height where rip channels develop. Lateral variation in wave height can be caused by offshore bathymetric changes that lead to areas of wave convergence (high waves) and divergence (low waves), as is the case for the rip currents found at La Jolla, California, where rip currents are developed by a longshore variation of wave breaker heights caused by wave refraction over offshore submarine canyons (Shepard and Inman, 1950).

In areas with relative uniform offshore bathymetry (as on the south Atlantic Bight and in the Carolinas), variations in wave height can be caused by alongshore variability of the nearshore bar, by superposition of different wave trains coming from different directions (Dalrymple, 1975; Dusek et al., 2010, Chapter 4, this volume), or by wave group forcing (Long and Ozkan-Haller, 2009). Mei and Liu (1977) demonstrated analytically how alongshore varying surf zone bathymetry could cause longshore pressure gradients. These types of rip currents are common at barred beaches where channels interrupt the continuity of the alongshore bar. Additionally, rip current circulations can be created through a feedback mechanism in which the initial wave height variation causes an incipient rip current to form that in turn interacts with the wave field, feeding more energy into the circulation system (Iawata, 1976; Murray et al., 2003).

Independently of the initial conditions, it is generally agreed that rip currents are maintained and persist because of topographic depressions (rip channels) within the surf zone. However, because the locations of the rip channels vary in space and time, it has been very difficult to obtain accurate field data. Haller et al. (2002)

presented experimental laboratory data of nearshore circulation on a barred beach with rip channels that showed that gaps in longshore uniform bars dominated the nearshore circulation system for both normal and obliquely incident waves. The SHORECIRC nearshore circulation model was used by Haas et al. (2003) to accurately simulate the rip currents observed by Haller et al. (2002), suggesting that SHORECIRC is an adequate tool to simulate bathymetrically controlled rip currents and other three-dimensional flows in the nearshore.

In realistic situations, the tide may modulate rip current activity directly by modulating local water depth and indirectly via modulation of the incident wave field. Brander and Short (2001) presented data from an Australian beach with a tidal range of 1.6 m and medium energy wave activity. They showed that rip current velocity is inversely proportional to water depth and is modulated by the tide, resulting in stronger rip current speeds at low tide conditions. Similar dependence of rip current velocity on tidal stage was shown by Brander (1999) who demonstrated that the velocity of rip is proportional to the cross-sectional area of the rip channel. Brander also noted that rip currents occur in intermediate type (Wright and Short, 1984) beaches, especially during the decreasing end of a highly energetic event. This dependence on rip currents and tidal stage is also reflected on the modified ECFLS LURCS index of Engle et al. (2002).

OBJECTIVES

The objective of this study was to design a physics-based nearshore hazard forecasting system to provide quantitative values of rip current aspects (current speed, relative density, and location offshore) that can be used with confidence to compute rip current risk indices by authorities charged with public safety. The methodology must be based on sound physical principles and be transferable to different locations so that it is adaptable for different morphologies found at different sites. It should be noted that the proposed prediction system does not allow predictions of exact locations and timing of rip currents. The only exception is the case of fixed bathymetry (areas near terminal groins and narrow passages through rocks). The aim is to provide quantitative predictions of potential rip current severity and thus promote better beach management and optimized deployment of professional rescuers.

This methodology aims at: (1) providing quantitative values of risks (current speed, relative location offshore, density of rip currents, etc.) in three dimensions; (2) accounting for the specificities of a particular site as expressed through historic morphology; and (3) providing solid physical parameters that can be incorporated into a more comprehensive risk index that could include sociological parameters (beachgoer attitudes, levels, swimming competence). Finally, this information can hopefully be included as part of a more comprehensive decision management tool operated by experts in beach management and sea rescue.

METHODOLOGY

The methodology for a rip current prediction system consists of the application of a suite of numerical models and appropriate bathymetric and offshore boundary conditions described in the following sections. In particular, a coupled 3-D circulation and wave propagation model is used to predict nearshore circulation and assess the potential for rip current development. The biggest unknown in the development of the prediction system is nearshore bathymetry and a methodology for dealing with this unknown is presented.

CIRCULATION MODELING

ROMS is a 3-D, free surface, bathymetry-following numerical model that solves finite difference approximations of Reynolds-averaged Navier-Stokes (RANS) equations using hydrostatic and Boussinesq approximations with a split-explicit time stepping algorithm (Shchepetkin and McWilliams, 2005; 2009). ROMS provides choices of model components like advection schemes (second, third, and fourth order), turbulence closure models (generic length scale mixing, Mellor-Yamada, Brunt-Väisälä frequency mixing, user-provided analytical expressions, K profile parameterization), boundary conditions, and others.

ROMS has undergone modifications that incorporate the methods described in Mellor (2003 and 2005) for use in nearshore environments. Haas and Warner (2009) compared the performance of ROMS to a quasi-3-D model (SHORECIRC) for obliquely incident wave-induced currents on a gently sloping beach and also for generation of rip currents. The wave information required for calculating wave-induced radiation stress was obtained by coupling ROMS to SWAN (obliquely incident wave case) and to a monochromatic wave driver REF/DIF (rip current circulation case). Following the remarks of Ardhuin et al. (2008) on Mellor's (2003 and 2005) implementation of depth-dependent radiation stress equations, Mellor (2008) produced an updated formulation coded into ROMS and evaluated against analytical solutions for rip current flows by Kumar et al. (2010). This updated coupled model also has been modified to include a vertical distribution of stresses that is more appropriate for sigma coordinates in very shallow waters (Kumar, 2010; Kumar et al., 2010). The updated model was tested through simulations of several cases that include: (a) obliquely incident spectral waves on a planar beach; (b) alongshore variable offshore wave forcing on a planar beach; (c) alongshore varying bathymetry with constant offshore wave forcing; and (d) nearshore barred morphology with rip channels. A number of quantitative and qualitative comparisons to previous analytical, numerical and laboratory data showed that the updated model more accurately replicates surf zone recirculation patterns (onshore drift at the surface and undertow at the bottom) as compared to the previous versions.

WAVE PROPAGATION MODEL (SWAN)

Several numerical models are available to describe wave transformation across arbitrary bathymetry, including phase-resolving and phase-averaged models. Phase resolving models can accurately predict free surface displacement due to wave field, but are computationally demanding and not practical for forecasting scenarios. Phase-averaged models that compute wave statistics such as height and period are more practical for forecasting. Monochromatic phase-averaged models typically overestimate focusing due to bathymetric rises, while spectral models are much less sensitive in this regard. A spectral model including wind forcing is preferred in some locations because the wind can strongly influence nearshore wave conditions as is the case for the Carolinas and the east coast in general (Austin and Lentz, 1999). The SWAN model (Booij et al., 1999; Ris et al., 1999) was chosen to meet project needs. Input to the model consists of a bathymetric grid, description of incident wave conditions (measured spectra in this application), and wind and mean current velocity fields.

Output provides comprehensive parameters of significant wave height, mean wave length, peak and mean surface period, and mean bottom period throughout the model domain. The model simulates shoaling, wave refraction due to bathymetric features and mean currents, energy input from winds, and energy loss arising from white capping, bottom friction, and breaking. This model is comprehensive, widely used, and compares favorably to a variety of data sets and analytical solutions. The same model has been set up to predict nearshore waves for the WFOs of Newport, Morehead City, and Wilmington, North Carolina (Willis et al., 2010) and will be implemented in Florida by the Melbourne WFO (Lascody, personal communication). Wave set up and non-stationary boundary conditions are included in the model.

BATHYMETRIC DOMAIN

The accuracy of any numerical modeling application can depend on the accuracy of the boundary conditions, initial conditions, forcings, and the bathymetry used to construct the grid domain. Bathymetry can change rapidly, especially in the very shallow nearshore waters. In fixed coastline areas (rocky coasts, jetties, concrete piers), the bathymetric domain can be considered constant and may be defined easily from existing charts or a bathymetric survey. However, in natural beach environments, the morphology can change rapidly in response to wave forcing. Although emerging technologies such as microwave radars (McNinch, 2006) may be able to provide remote assessment of nearshore beach morphologies, they are still experimental and not available for routine use. However, a rigorous analysis of beach profile data routinely collected in the course of beach monitoring projects can be incorporated in the design of synthetic model domains suitable for a particular location. For barred morphologies, the bathymetric problem requires definition of the mean bathymetric profile, locations and dimensions of alongshore bars, rip channel widths, and rip channel spacing.

In the approach presented here, a synthetic bathymetric set-up is based on local characteristics of the region of interest. Historical nearshore surveys can be utilized to construct averaged and extreme rip-favorable beach profiles. Such survey data describe the profile shape across a beach out to a pre-determined water depth. Such profiles from each benchmark and location can be analyzed using empirical orthogonal function (EOF) analysis (Aubrey, 1979; Gao et al., 1998) as described below.

For a beach profile monitored n times at m cross-shore locations, a correlation matrix M $(m \times n)$ with the elements $h (x, t)$ can be created where x is the cross-shore location, t is the time, and h is the elevation above the sea bed. The two square matrices formed using M are:

$$A = \frac{1}{(m \times n)}\left(MM^T\right) \qquad B = \frac{1}{(m \times n)}\left(M^T M\right) \qquad (5.1)$$

where M^T is the transpose matrix of M. Matrix A has a dimension of $m \times m$ and possesses m eigenvalues (e) and eigenfunctions (λ) such that:

$$Ae_i = \lambda e_i \qquad (5.2)$$

and

$$Bc_j = \lambda c_j \qquad (5.3)$$

where c is a temporally related eigenfunction.

Aubrey (1979) has shown that only the highest three eigenvalues (λ_1, λ_2, λ_3) and their associated eigenfunctions (e_1, e_2, e_3; c_1, c_2, c_3) explain the majority (>90%) of the variance representing beach profiles. The bed elevation (h) at location x and time t can be re-constructed using:

$$h(x,t) = a_1 e_1(x,1) \times c_1(t,1) + a_2 e_2(x,2) \times c_2(t,2) + a_3 e_3(x,3) \times c_3(t,3) \qquad (5.4)$$

where α is the normalizing factor $a = (\lambda mn)^{0.5}$. In this chapter, only spatial eigenfunctions (e_1, e_2 and e_3) are utilized. The first eigenfunction (e_1) from the analysis is interpreted to represent the mean profile; the second (e_2) and third (e_3) eigenfunctions are correlated to the bar topography. The second eigenfunction typically shows a large maximum correlated to the location of the summer berm and a minimum in the area of the offshore bar. The third eigenfunction shows a broad maximum correlated to the location of the low-tide terrace.

Thus the analysis of historical data can reveal a mean beach profile and locations and dimensions of bars. Synthesis of data collected from field studies of rip currents indicates a distinct scaling of rip channel spacing (L) with surf zone width (X_c) and channel width (L_t), respectively. The ratio of channel spacing over surf zone width was found to be 2.7 to 4 and 1.5 to 8 by Haller et al. (2002) and Huntley et al. (1992), respectively. Rip spacing appears to be five times the channel width (Haller et al., 2002, Brander and Short, 2001, Aagaard et al., 1997).

Scaling a nearshore bathymetric model domain can be developed by using the following composite beach profile configurations: (1) alongshore uniform beach morphology equivalent to the mean profile (open lateral boundary conditions); (2) alongshore uniform beach profile consisting of the mean profile plus the bar with a height as revealed by EOF analysis (open lateral boundary conditions); (3) alongshore variable beach profile consisting as in (2) with a channel interrupting the bar to simulate the potential of rip current development due to bathymetric features. The latter presents the most rip-favorable condition and accurate simulation of the flows for such conditions can indicate the risk of rip current development. The scaling presented in the previous paragraph can be used to define the 3-D bathymetric domain.

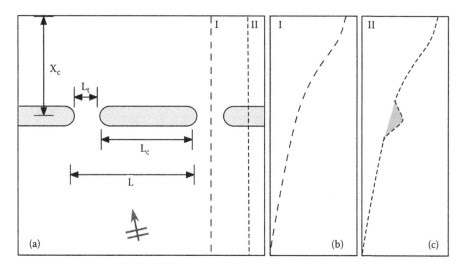

FIGURE 5.1 Length scales required to construct a bar trough morphology domain to model rip current channel: (a) plan view; (b) and (c) beach profiles at rip channel and over bar, respectively. Bar length (L) is 3.3 times the width (X_c) of the surf zone that roughly corresponds to the distance between the coastline and offshore location of the bar. The latter is estimated from empirical orthogonal function analysis of beach profile data. Rip channel width (L_t) is scaled with bar length ($L_t = 1/5$).

Figure 5.1 shows an experimental design for the domain consisting of a barred beach incised with two rip channels. The profile for the barred beach is obtained by superposition of bar information on the mean profile as in (II) above. The profile for the rip channel is created by joining a straight line from the bar trough to the starting or end point of the bar crest (see Figure 5.2c for details). The rip channel spacing (L) and channel width (L_t) are calculated by utilizing the ratio with surf zone width (X_c), as shown by previous field and laboratory studies. The transition from the rip channel to barred beach is achieved via a sinusoidal function.

STUDY CASE: DUCK, NORTH CAROLINA

As a demonstration of the proposed methodology, we present a study that applies to the bathymetry of Duck. The procedure for the development of a rip current prediction tool consists of four distinct steps: (1) development of the bathymetric domain; (2) set-up of the numerical domain; (3) numerical simulations; and (4) rip current condition prediction for different waves and/or tidal stages.

DEVELOPMENT OF MORPHOLOGY

Beach profiles covering a period of 20 years were analyzed via EOF. The data were collected monthly by the Field Research Facility at Duck (http://www.frf.usace.army. mil/bathy-main.shtml). EOF analysis led to creation of a beach profile decomposed into a mean [h(x)] and a bar [h1(x)] function. Figure 5.2a shows EOF 1, the mean

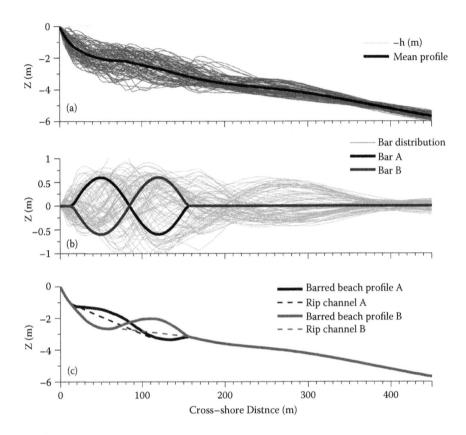

FIGURE 5.2 Empirical orthogonal function analysis of 20-year long monthly beach profile record at Duck, North Carolina, and creation of a synthetic profile: (a) individual (grey lines) and mean (black line) profiles; (b) cross-shore profiles of bar functions obtained from analysis of individual profiles (solid and dashed lines show two most common bar distributions); (c) solid black and grey lines correspond to barred beach profiles obtained by adding mean profile in (a) to bar distribution in (b). The dashed black and grey lines show the rip channel profile obtained by joining a straight line from the bar trough to the end or beginning of the bar crest for Cases A and B, respectively.

profile for entire data set. The second and third EOFs (Figure 5.2b) represent the bar and terrace functions, respectively. These two EOFs explain 60.6 and 39.3% of the variability found around the mean profile and are indicative of the location, width, and height of the bar that develops in the study area. The bar width was defined as the cross-shore distance from bar trough to bar crest.

Identification of the maximum value and its location along the cross-shore profile was used to identify the bar height and location for each time measured. The most common bar distribution pattern can be superimposed on the mean profile to create a possible barred beach and further modified to obtain rip channel profile configurations (Figure 5.2c). A statistical analysis in the form of a joint probability plot of bar height versus width, bar distance from shoreline versus bar width, and bar distance

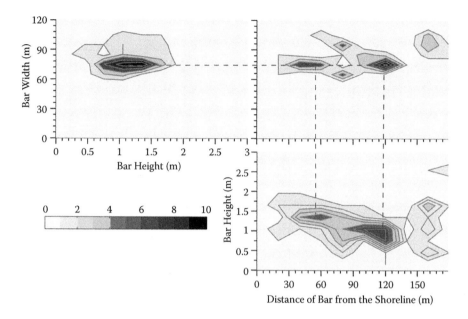

FIGURE 5.3 Joint probability distribution of bar characteristics (width, height, distance from shoreline) for Duck as identified by EOF analysis (see Figure 5.2). Note two distinct bars that correspond to typically observed summer and winter profiles.

from shoreline versus bar height was used to identify the most common bar configurations developing at the study site. This analysis incorporates local information such as sediment size and local wave climate. The longer the time covered by the data base, the more reliable and significant the correlations.

Figure 5.3 shows the joint distribution analysis for Duck. Two types of bars are the most common and, not surprisingly, correspond to the development of the so-called summer and winter profiles, respectively. Quantitatively, these conditions are: (a) bar located 60 m from the shoreline, 1.2 m high and 70 m wide; (b) bar located 120 m from the shoreline, with a height of 1.0 m and width of 70 m. Using the scaling analysis discussed in the previous section, the channel width and spacing values are 40 m and 200 m for (a) and 80 m and 400 m for (b).

These bar characteristics can be superimposed on the mean beach profile (see Figures 5.1 and 5.2) and used to develop the bathymetric domain shown in Figure 5.4 for two bar cases presented in the next section (hereafter called Cases A and B, respectively).

Model Set-Up

For each condition, a model domain is constructed with a grid resolution 4 m × 4 m in the cross-shore and alongshore directions, respectively. The set-up of case study includes incidence of alongshore uniform wave height on the bathymetric domain.

The domain for Case A is 500 m in the cross-shore and 640 m in the alongshore direction, while for Case B it is 500 m in the cross-shore and 800 m in the alongshore directions. Vertically, the domain is distributed in ten equally distributed

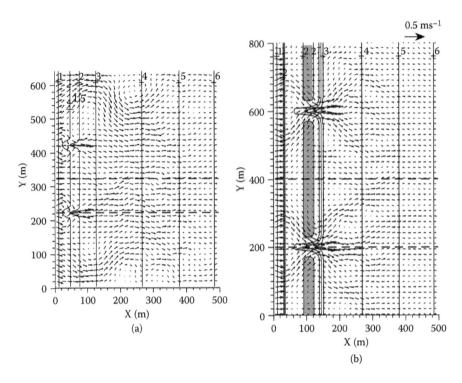

FIGURE 5.4 Synthetic bar trough morphology for Cases A (left panel) and B (right panel) and corresponding depth-averaged current vectors (black arrows) for a 2-m height and 8-sec period wave, approaching the coastline at 0 degrees. Solid black lines indicate bathymetric contours. Locations of transects at which the vertical profiles of cross-shore velocity are shown in Figures 5.5 through 5.7 are indicated by dashed grey lines.

sigma layers. Closed boundary conditions are used at the two lateral sides and the shoreline, while gradient (Neumann) boundary conditions were used at the offshore boundary. A logarithmic bottom friction with a roughness length of 0.005 m—a value close to those reported from field studies (Feddersen et al., 1998)—was used.

The wave model (SWAN) was run for the same grid as ROMS. The wave forcing applied at the offshore boundary was directed perpendicular to the domain, had a period of 8 sec and a significant wave height of 2 m. The selected period corresponds to the most common period at the site as revealed through a joint probability analysis of wave data from the site. The corresponding most common wave height is 1 m. However, in this test case, the wave height was increased to represent increased wave conditions. The wave forcing is described by a directional spectrum consisting of 20 frequency bands in the range 0.04 to 1 Hz, and 36 directional bins of 10 degrees each from 0 to 360 with a directional spreading of 6 degrees. The bottom friction used in SWAN is based on the eddy viscosity model of Madsen et al. (1988) with a bottom roughness value of 0.05 m. The modeling system for this case was configured in two-way coupling mode where exchange of wave and current information takes place between ROMS and SWAN at a synchronization interval of 15 sec. The coupled model system was run for a simulation time of 2 hr over which the computational domain achieves stability.

RESULTS

The depth-averaged velocities for each case are shown in Figure 5.4 for a simulation with a mean tidal water level. It is characteristic that for each case a rip cell develops and occupies the region between the bar and the shoreline, with the main rip current developing over the rip channel location. Subtle differences are observed between the two cases. In Case A, the offshore-directed current develops exactly at the location of the rip channel (some 50 m from the shoreline), while in Case B, offshore-directed flows start developing well inshore of the rip channel and intensify at the location of the channel. In both cases, the depth-averaged offshore-directed current is approximately 0.5 m/sec.

The vertical distribution of the cross-shore current under the same conditions is shown in Figure 5.5 for locations over the bar and over the channel for Cases A and B, respectively. It is characteristic that the flow over the bar in both cases is much smaller than through the channel. Maximum offshore velocities occur in the channel and some 1 m and 1.5 m below the sea surface for Cases A and B, respectively.

The effect of tidal variability is shown through the comparison of simulations at mean, high, and low water levels for the same offshore wave conditions (Figures 5.6 and 5.7) in Cases A and B. In Case A, the rip current strength at high, mean, and low water levels are 0.47 ms^{-1}, 0.61 ms^{-1}, and 0.66 ms^{-1}, respectively. For Case B, these velocities are 0.49 ms^{-1}, 0.58 ms^{-1}, and 0.65 ms^{-1}. Characteristically, the rip current strength increases during the low water conditions, exhibiting a clear dependence of wave conditions on tidal stage.

DISCUSSION

The model has been shown to predict rip current conditions for a normal wave incidence. Due to space limitations, no examples were given for waves approaching from different angles or sensitivity of the results on the parameterizations used for the development of the bathymetric domain.

Kumar et al. (2010) presented results on the strength of rip currents as a function of wave angle of approach. The development of rip current circulation on an alongshore bar trough morphology domain was subjected to offshore waves with height of 0.5 m, period of 3 sec, and incident at angles of 0, 5, 10, and 20 degrees with respect to shore-normal. The results showed that as the wave incidence angle increases from 0 to 20 degrees, the angle of exit of the rip current increases with respect to shore-normal. The trend is linear and for angles exceeding 20 degrees, no rip current is observed as all flow becomes almost parallel to the shoreline. As expected, the strength of alongshore velocity increased as the wave angle of incidence increased, making longshore current a potential hazard in the nearshore.

For normal wave incidence, primary and secondary circulation cell formation occurs outside the rip channel and close to the shoreline, respectively. The cells are symmetric about the rip channel center with opposite signs of vorticity, indicating a

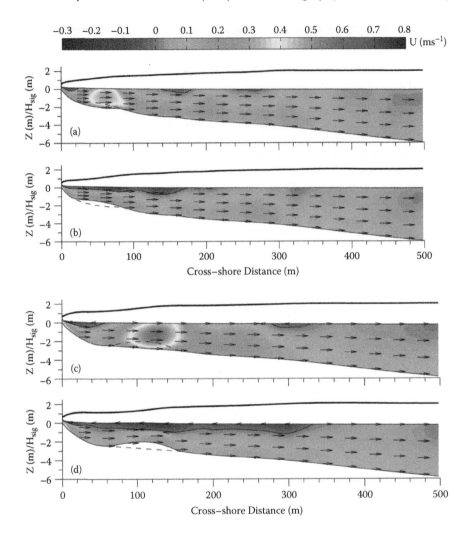

FIGURE 5.5 (*See color insert.*) Vertical distribution of cross-shore velocity at the center of a rip channel (a and c) and bar (b and d) for bathymetric domains corresponding to Cases A (a and b) and B (c and d), respectively. Thick black lines show cross-shore profile of significant wave height. Locations of transects are shown in Figure 5.4.

reverse sense of circulation. With increasing angle of wave incidence, the secondary circulation pattern weakens, but the primary circulation pattern is reinforced. At a wave incidence of 10 degrees, the secondary circulation cell close to the shoreline disappears, while the 20-degree incidence shows only one circulation cell constrained at the original location where primary circulation was observed.

Kumar et al. (2010) showed that rip current velocity at these locations was generally stronger when wave incidence was at 5 and 10 degrees. Onshore of the channel, maximum offshore-directed flow within the channel area occurs at 5 degrees; at

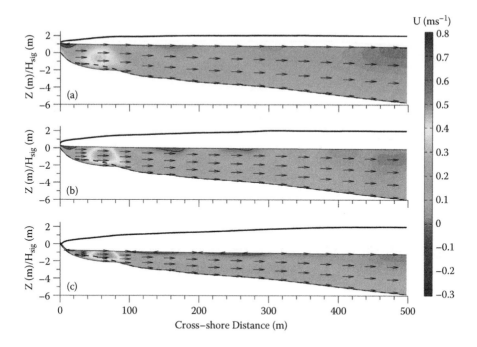

FIGURE 5.6 (*See color insert.*) Case A: Cross-shore velocity (color contours) and significant wave height (black line) distribution along a transect through the rip channel location for three tidal stages: (a) high; (b) mean; and (c) low water levels. Thin black lines represent wave height distribution across the domain as estimated by SWAN. Note how the speed of the rip current increases with decreasing water level, while the offshore location of the maximum rip speed remains the same. The vectors indicate direction of flow; current strength is shown by the color-filled contours.

transects within the channel, rip current velocity is slightly higher for 10 degrees in comparison to 5 degrees incidence. Higher angles of incidence (> 20 degrees) inhibit rip currents due to inertia of alongshore flow. Aagard et al. (1997) observed similar increases in the rip current velocity due to oblique incidence, while Haller et al. (2002) observed an abrupt increase in cross-shore velocity for wave incidence angle of 10 degrees. This behavior is probably due to an increase in alongshore radiation stress created by breaking of obliquely incident waves at the bar crest.

In an attempt to examine the sensitivity of the model to the parameterization of the bar trough morphology, runs for case A were undertaken after changing the channel width and holding all other parameters constant. The results are shown in Figure 5.8. It is characteristic that the circulation cell that develops increases with increased channel width, implying that more water mass exits through the rip channel. However, since the channel width also increased, it appears that the increased mass of water is trying to exit through an increased cross-sectional area so that the net rip current speed seems to be the same (approximately 0.45m/sec), independently of the channel width selected.

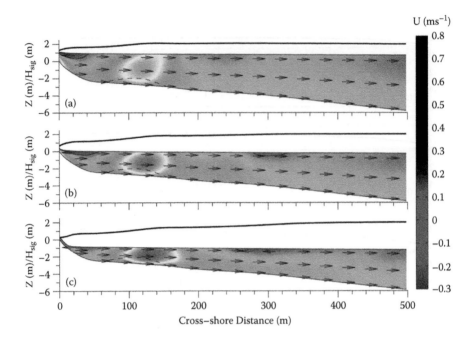

FIGURE 5.7 (*See color insert.*) Case B: Cross-shore velocity (color contours) and significant wave height (black line) distribution along a transect through the rip channel location for three tidal stages: (a) high; (b) mean; and (c) low water levels. Thin black lines represent wave height distribution across the domain as estimated by SWAN. Note how the speed of the rip current increases with decreasing water level, while the offshore location of the maximum rip speed remains the same. The vectors indicate direction of flow; current strength is shown by the color-filled contours. The relative location of the bar is shown as a grey line.

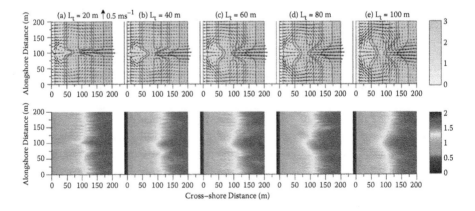

FIGURE 5.8 (*See color insert.*) Circulation (depth-averaged current vector, top row) and significant wave height distribution (bottom row) for bathymetric domain of Case B with varying channel width (L_t) of (a) 20 m; (b) 40 m; (c) 60 m; (d) 80 m; and (e) 100 m.

CONCLUSIONS

A methodology has been developed to utilize existing historical bathymetric surveys and construct realistically based hypothetical bathymetric data sets to represent local bar trough morphology. The constructed bathymetry provides several scenarios of realistic cases that can be simulated using a public domain, well-tested, physics-based, full 3-D circulation model (ROMS) with a wave propagation model (SWAN) to nowcast and/or forecast surf zone conditions (development of rip currents and actual and expected current strengths). Such approaches can be implemented using existing wave and current forecasting systems for boundary conditions; bottom morphology created from historical beach profile data or remotely sensed morphology (x-band radar, images, etc.). The derived predictions are based on runs for different variation of morphology as defined by historical analysis; the highest rip current speeds would be the most useful for beach management decisions.

An advantage of this approach is that the methodology incorporates site-specific characteristics to provide local characterization of flows. However, the authors recognize and caution that knowing the strength of rip currents does not protect a community from the hazards of rip currents. Instead the goal is for this information to be provided to beach management personnel as a comprehensive decision-making tool that includes other parameters such as sociological and behavioral patterns, beach safety personnel availability, and other factors. One way that this tool can contribute significantly is by providing a priori knowledge of conditions to be expected to assist with planning and optimized deployment of available resources. A period of intense evaluation would be useful prior to implementation of this method as an operational tool.

ACKNOWLEDGMENTS

Voulgaris and Kumar were supported by a NOAA/IOOS grant (Integration of Coastal Observations and Assets in the Carolinas in Support of Regional Coastal Ocean Observation System Development in the Southeast Atlantic) and a cooperative agreement between the U.S. Geological Survey and the University of South Carolina as part of the Carolinas Coastal Change Processes Project. In addition, Voulgaris was partially supported by the National Science Foundation (Awards: OCE-0451989 and OCE-0535893). Finally, we would like to thank the Field Research Facility of Duck, North Carolina for maintaining a free and easy-to-use database and also the ROMS and SWAN code developers for their hard work in developing and maintaining these open source codes.

REFERENCES

Aagaard, T., B. Greenwood, and J. Nielsen. 1997. Mean currents and sediment transport in a rip channel. *Marine Geol.,* 140: 25–45.

Allard, R., J. Dykes, Y.L. Hsua et al. 2008. A real-time nearshore wave and current prediction system. *J. Marine Syst.,* 69: 37–58.

Ardhuin, F., A.D. Jenkins, and K.A. Belibassakis. 2008. Comments on "three-dimensional current and surface wave equations." *J. Phys. Oceanogr.,* 38: 1340–1350.

Aubrey, D.G. 1979. Seasonal patterns of onshore/offshore sediment movement. *J. Geophys. Res.*, 84: 6347–6354.

Austin, J. and S.J. Lentz. 1999. Relationship between synoptic weather systems and meteorological forcing on the North Carolina inner shelf. *J. Geophys. Res.*, 104: 18159–18185.

Booij, N., R.C. Ris, and L.H. Holthuijsen. 1999. Third generation wave model for coastal regions I. Model description and validation. *J. Geophys. Res.*, C4: 7649–7666.

Brander, R.W. 1999. Field observations on the morphodynamic evolution of low energy rip current system. *Marine Geol.*, 157: 199–217.

Brander, R. and W. Short. 2001. Flow kinematics of low-energy rip current systems. *J. Coastal Res.*, 17: 468–481.

Bowen, A.J. 1969. Rip currents I. Theoretical investigations. *J. Geophys. Res.*, 74: 5467–5478.

Dalrymple, R.A. 1978. Rip currents and their causes. *Proc. 16th Intl. Conf. on Coastal Engineering.* ASCE: 1414–1427.

Devaliere, E., J. Hanson, and R. Luettich. 2009. High resolution nearshore wave model for the Mid-Atlantic Coast. U.S. Army Corps of Engineers Field Research Project S07-66810, Final Report.

Dusek, G., H. Seim, J. Hanson et al. 2010. Analysis of rip current rescues at Kill Devil Hills, North Carolina. Chapter 4, this volume.

Ebersole, B. and R.A. Dalrymple. 1980. Numerical modeling of nearshore circulation. *Proc. 17th Intl. Conf. on Coastal Engineering.* ASCE: 2710–2725.

Engle, J., J. MacMahan, R.J. Thieke et al. 2002. Formulation of a rip current predictive index using rescue data. *Proc. Natl. Conf. on Beach Preservation Technology.* FSBA: Biloxi. MS.

Gao, S., M. Collins, and J. Cross. 1998. Equilibrium coastal profiles II. Evidence from EOF analysis. *Chin. J. Oceanol. Limnol.*, 16: 193–204.

Haller, M.C., R.A. Dalrymple, and I.A. Svendsen. 2002. Experimental study of nearshore dynamics on a barred beach with rip channels. *J. Geophys. Res.*, 107: doi 10.1029/2001 JC000955.

Haller, M.C. and R.A. Darlymple. 2001. Rip current instabilities. *J. Fluid Mech.*, 433: 161–192.

Haas, K.A., I.A. Svendsen, M.C. Haller et al. 2003. Quasi three-dimensional modeling of rip current systems. *J. Geophys. Res.*, 108: doi 10.1029/2002 JC001355.

Haas, K.A. and J.C. Warner. 2009. Comparing a quasi-3D to a full 3D nearshore circulation model: SHORECIRC and ROMS. *Ocean Model.*, 26: 91–103.

Iawata, N. 1976. Rip current spacing. *J. Oceanogr. Soc. Jpn.*, 32: 1–10.

Kinsman, B. 1965. *Wind Waves.* Prentice-Hall, New York.

Kirby, J.T. and R.A. Dalrymple. 1982. *Numerical Modeling of the Nearshore Region.* Research Report CE-82-24. Ocean Engineering Program, University of Delaware, Newark.

Kumar, N., G. Voulgaris, and J.C. Warner. 2010. Implementation of an updated radiation stress formulation and applications to nearshore circulation. *Coastal Eng.* (submitted).

Lascody, L.L. 1998. East Central Florida rip current program. *Natl. Weather Dig.*, 22.

Lavrenov, I.V. 2003. *Wind Waves in Oceans.* Springer, Heidelberg.

Longuet-Higgins, M.S. 1970a. Longshore currents generated by obliquely incident sea waves 1. *J. Geophys. Res.*, 75: 6778–6789.

Longuet-Higgins, M.S. 1970b. Longshore currents generated by obliquely incident sea waves 2. *J. Geophys. Res.*, 75: 6778–6789.

Longuet-Higgins, M.S. and R.W. Stewart. 1964. Radiation stress in water waves: a physical discussion with applications. *Deep Sea Res.*, 11: 529–562.

Lushine, J.B. 1991. A study of rip current drownings and related weather factors. *Natl. Weather Dig.*, 16.

McNinch, J.E. 2006. Bar and swash imaging radar (BASIR): a mobile x-band radar designed for mapping sand bars and swash-defined shorelines over large distances. *J. Coastal Res.*, 23.

Murray, A.B., M. LeBars, and C. Guillon. 2003. Tests of a new hypothesis for non-bathymetrically driven rip currents. *J. Coastal Res.*, 19: 269–277.

Mei, C.C. and P.L. Liu. 1977. Effects of topography on the circulation in and near the surf zone: linearized theory. *J. Estuar. Coastal Marine Sci.*, 5: 25–37.

Mellor, G.L. 2003. The three-dimensional current and surface wave equations. *J. Phys. Oceanogr.*, 33: 1978–1989.

Mellor, G.L. 2005. Some consequences of the three-dimensional currents and surface wave equations. *J. Phys. Oceanogr.*, 35: 2291–2298.

Mellor, G.L. 2008. The depth-dependent current and wave interaction equations: a revision. *J. Phys. Oceanogr.*, 38: 2587–2596.

NOAA. July 16, 2004. National Weather Service Instruction 10-310. Marine and Coastal Weather Service Program. NWSPD10-3 (http://www.nws.noaa.gov/directives).

Ris, R.C., N. Booij, and L.H. Holthuijsen. 1999. A third-generation wave model for coastal regions II. Verification. *J. Geophys. Res.* C4: 7667–7681.

Shchepetkin, A.F. and J.C. McWilliams. 2005. The regional oceanic modeling system: a split-explicit, free surface, topography-following coordinate oceanic model. *Ocean Model.*, 9: 347–404; doi:10.1016/j.ocemod.2004.08.002.

Shchepetkin, A.F. and J.C. McWilliams. 2009. Correction and commentary for ocean forecasting in terrain-following coordinates: formulation and skill assessment of the regional ocean modeling system. *J. Comput. Phys.*, 228: 8985–9000.

Shepard, F.P. and D.L. Inman. 1950. Nearshore Water Circulation Related to Bottom Topography and Wave Refraction. *Eos Trans. AGU*, 31: 196–213.

Shi, F., I.A. Svendsen. J.T. Kirby et al. 2003. A curvilinear version of a quasi-3D nearshore circulation model. *Coastal Eng.*, 49: 99–124.

Short, A.D. 1985. Rip current type. spacing and persistence: Narrabeen Beach. Australia. *Marine Geol.*, 65: 47–71.

Sonu, C.J. 1972. Field observation of nearshore circulation and meandering currents. *J. Geophys. Res.*, 77: 3232–3247.

Svendsen, I.A., K. Haas, and Q. Zhao. 2002. *Quasi-3D Nearshore Circulation Model SHORECIRC*. Version 2.0'. Report CACR-02-01. Center for Applied Coastal Research, University of Delaware, Newark.

Thornton, E.B. 1970. Variations of longshore currents across the surf zone. *Proc. 12th Intl. Conf. on Coastal Engineering*, ASCE, 291–308.

Voulgaris. G., J.C. Warner, P.A. Work et al. 2004. The South Carolina Coastal Erosion Study: integrated circulation and sediment transport studies: project overview. *Eos Trans. AGU. Fall Meeting Suppl.*, Abstract OS21B-1224.

Voulgaris, G. and M.B. Collins. 2000. Sediment resuspension on beaches: response to breaking waves. *Marine Geol.*, 167: 167–187.

Voulgaris, G., M.D. Simmonds, H. Howa et al. 1998. Measuring and modelling sediment transport on a macrotidal ridge and runnel beach. *J. Coastal Res.*, 14: 315–330.

Voulgaris, G., T. Mason, and M.B. Collins. 1996. The energetics approach for suspended sand transport in macrotidal ridge and runnel beaches. *Proc. 25th Intl. Conf. on Coastal Engineering*. ASCE, 3948–3961.

Warner, J.C., C. Sullivan, G. Voulgaris et al. 2004. The South Carolina Coastal Erosion Study: numerical modeling of circulation and sediment transport in Long Bay. *Eos Trans. AGU. Fall Meeting Suppl.*, Abstract OS21B-1226.

Willis, M.C., E. Devaliere, J. Hanson et al. 2010. Implementing the SWAN wave model at three East Coast National Weather Service Offices. 14th Symp. on Integrated Observing and Assimilation Systems for the Atmosphere. Oceans. and Land Surface, Atlanta, Extended Abstract. 5B.7.(http://ams.confex.com/ams/90annual/techprogram/paper_164982.htm).

Wright, L.D. and A.D. Short. 1984. Morphodynamic variability of surf zone and beaches. *Marine Geol.*, 56: 93–118.

6 Surf Zone Hazards
Rip Currents and Waves

Robert G. Dean and R. J. Thieke

CONTENTS

INTRODUCTION

Rip current and other surf zone hazards can be reduced through better knowledge of a threat level at a given time and location and by improving the response capabilities of individuals at risk. The presence of lifeguards substantially reduces such risks, but lifeguards cannot be present at all locations where swimmers and surf zone hazards coexist. Present approaches to quantify these threats involve broad-based prediction methods and a review of the rip current history at a particular site; swimmer education could presumably reduce the threat. This chapter focuses on several approaches to reducing rip currents hazards:

- Rational accounting for the quantitative effects of bar height
- Improved understanding of the stability of individuals against currents and waves in the surf zone
- Better detection of rip currents for daily quantification and calibration of rip current prediction systems
- Fine tuning of existing broad-based prediction systems
- More effective application of present and future advisory systems
- Emphasizing personal responsibility for water safety

BETTER RIP CURRENT PREDICTIVE MODELS

ROLE OF TIDES IN THREE EXISTING PREDICTION METHODS

Valuable rip current predictive models were first developed by Lushine (1991) and modifications for local application and incorporation of more complete wave information were introduced by Lascody (1998) and Engle (2003), respectively. The essential ingredients of these models include weighting factors for wave height, winds, tides, and rip current persistence. The role of tides leading to a suggested additional parameter is examined below.

All three rip current indices discussed above recognize the role of reduced tidal stage to some degree in increasing rip current hazards. Lushine (1991) developed the LURCS (LUshine Rip Current Scale) index defined by six categories ranging from 0 to 5 for the southeast Florida coast. He incorporated a "tidal factor" that increased the risk by one category if a tide was within the range of –2 to +4 hours of low tide. An increase of one category raises the risk by an amount that depends on the risk level without the tidal factor. To illustrate, a Category 1 risk level is defined as "caution for weak or non-swimmers; weak rip currents possible." An increase to Category 2 at or near low tide results in an elevation of the rip current threat to "caution for all; moderate rip currents possible."

Lascody (1998) adapted the LURCS method to be more applicable for the east central Florida coast. The main reason for this adaptation was to account for observed risk due to long period ocean swells prevalent in that area rather than onshore winds determined by Lushine to be primary risk factors in southeast Florida. Rather than an explicit timing of low tide as a factor in the hazard level, Lascody's method adds 0.5 to a level if the "astronomical tides are higher than normal (i.e., near full moon)." This increase is compared, for example, to a total rip current threat level of 4.5 to 5.0 for which the hazard is "much greater than normal." Thus in the Lascody method, tides also contribute significantly to rip current threat levels.

Engle (2003) developed a rip current index addressing winds, tides, and wave factors and also accounted for the width of the wave spectrum, with greater rip current threats associated with narrower wave spectra. As for the Lascody method, Engle's method was developed and calibrated for the Volusia County area in east central

Florida. A factor of 2 is added in the Engle method if the tide is lower than −0.75 to −0.50 m. A rip current warning is issued for a level >5 and a very high threat warning for a level >9. Tides are recognized as significant contributors to rip current threats.

Based on the several types of rip currents (Dalrymple, 1978), the following section focuses on the type of rip current fixed by a channel through an offshore bar.

Processes Associated with Formation of Rip Current Channels

It is useful to first discuss the competing forces associated with rip channel formation. Waves propagating over a submerged nearshore bar result in a flow of water into the region landward of the bar. For purely two-dimensional conditions, this water "ponds" landward of the bar, resulting in a super-elevation of water in the surf zone that is hydraulically responsible for the seaward return of this water. Waves tend to mobilize sand so that it tends to flow down slope under the mobilizing effects of waves and the action of gravity, contributing to closure of the channel. When a vestigial channel is formed through a bar, the seaward flowing currents maintain the channel and the sand adjacent to the channel mobilized by the waves tends to flow into and close the channel. Although it is not possible to quantify precisely each of these competing forces, it is possible to develop simple relationships that illustrate their relative roles.

One possible scenario for the initiation of a rip channel is as follows. During a storm that includes high waves and possibly elevated water levels, a two-dimensional bar is formed at a distance from the shoreline consistent with the breaking waves and the local water depth. During or after the storm, with return to normal tide levels but with high waves, "ponded" water tends to seek and enlarge a low bar area. Infragravity waves associated with wave groups may contribute to this initiation process during storms whereby low water levels associated with large wave groups cause high ponded levels inside the surf zone and release of this ponded water between groups of high waves.

The presence of a bar enhances the probability of rip current formation and maintenance. Further, it follows that the proportion of the water depth occupied by the bar is likely relevant to the magnitude of the rip current threat. For example, a bar that occupies 50% of the water column is more likely to form and maintain a rip current than a bar that occupies, say, 10% of the column. To the best of our knowledge, this proportionality factor has not been included explicitly in previous rip current prediction methods, in part perhaps, because the developers of the indices did not find this data readily available. However, somewhat ironically, lifeguards are uniquely qualified to quantify such information on a daily or more frequent basis. Thus, it is recommended that this bar proportionality factor be tested as a relevant parameter in rip current prediction methods. The sections below consider the effects of relative bar height on rip channel formation and maintenance.

Rip Current Channel Formation

A very simple and heuristic model is based on linear shallow water wave theory and a simple channel maintenance criterion. Depending on its height, an incident wave

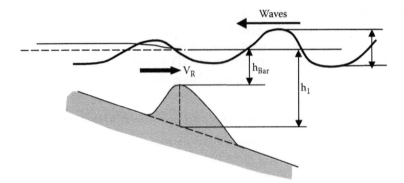

FIGURE 6.1 Simple two-dimensional model for increased water level landward of the bar inducing a return flow velocity V_R. The h_1 represents depth in the absence of the bar at the bar crest and was selected for illustrative purposes.

may break at any location seaward or landward of the bar. It will be shown, based on simple wave theory, that waves breaking seaward of the bar are more favorable to rip current channel formation and maintenance. For illustration purposes, we first consider the breaking to occur seaward of the bar crest at a depth h_1 equal to the depth if the bar were not present. Referring to Figure 6.1 representing a two-dimensional system, the volumetric landward water transport rate per unit of beach length q by the waves at a depth equal to the projected depth under the crest of the bar is:

$$q = \frac{E_1}{\rho C_1} \tag{6.1}$$

in which

$$E_1 = \frac{\rho g H_1^2}{8}$$

is the wave energy density at depth h_1 seaward of the bar, g is the acceleration of gravity, ρ is the mass density of water, and $C_1 = \sqrt{g h_1}$ is the wave celerity (speed) for depth h_1 at the shallow water asymptotic limit for linear waves. Applying a simple depth-limited breaking assumption, this can be reduced to:

$$q = \frac{\kappa^2}{8} \sqrt{g h_1^3} \tag{6.2}$$

in which κ is the ratio of breaking wave height to water depth and depends on beach slope; however, for steep slopes, it is on the order of 0.8. If the volumetric transport rate in (6.2) above flows seaward across the bar as a return flow, this return velocity V_R is approximately:

$$V_R = 0.08 \frac{\sqrt{g h_1^3}}{h_{Bar}} \tag{6.3}$$

Thus, the smaller the depth over the bar, the greater the return velocity for a given incident wave height. If the return velocity is sufficiently strong, presumably rip channels may form at certain locations along the bar. For the moment, this analysis is purely two-dimensional and no attempt to consider rip channels has been made. Denoting the tide level by η, the return velocity accounting for the presence of the tides is:

$$V_R = 0.08 \frac{\sqrt{g(h_1 + \eta)^3}}{h_{Bar} + \eta} \qquad (6.4)$$

To examine the above relationships further, four measured barred beach profiles are considered. The state of Florida maintains an extensive data base including shoreline positions and beach profiles at more than 3,900 locations around the state's 1,180-km predominantly sandy shoreline. Although the first available shoreline data are from the mid to late 1800s, the earliest available beach profiles appeared in the mid 1970s. Figure 6.2 shows four profiles at different times at a location in Palm Beach County. These profiles will be utilized to determine the effects of tides on return flow velocities.

Each of the four profiles was extended under the bars to determine an approximate depth h_1 (as shown in Figure 6.1) and the depth over the bar h_{Bar}. Each of these is related to NGVD which, for purposes here, is considered mean sea level. The spring and mean tidal ranges in the Palm Beach area are 0.95 and 0.79 m, respectively.

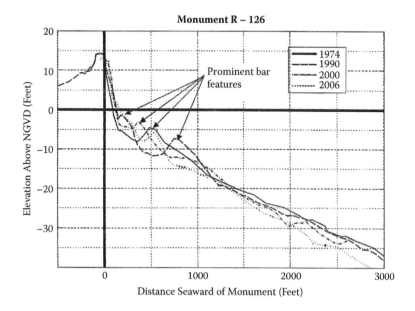

FIGURE 6.2 Profiles taken at Florida Department of Environmental Protection Monument R-126 in Palm Beach County on four dates. Note prominent bar features.

TABLE 6.1

Bar Characteristics for Four Palm Beach Profiles (Figure 6.2) and Return Flow Velocities for Spring and Mean High and Low Tides

Profile Date	h_1 (m)	h_{Bar} (m)	V_R (m/sec) Spring Tide High	V_R (m/sec) Spring Tide Low	V_R (m/sec) Mean Tide High	V_R (m/sec) Mean Tide Low
1974	2.74	1.46	0.75	0.87 (16%)	0.75	0.85 (13%)
1990	4.27	2.13	0.99	1.12 (13%)	1.00	1.10 (10%)
2000	2.13	0.91	0.76	1.24 (62%)	0.77	1.12 (45%)
2006	2.74	1.52	0.72	0.82 (14%)	0.73	0.80 (10%)

Note: Wave breaking in water depth (h_1 + tide level). Percentage values in return velocities are increased at low relative to high tides.

Equation (6.4) was applied to these measured bar scenarios to determine the return flows at high and low tides for spring and mean tidal ranges, considering the waves to break at h_1 and to govern the wave-induced volumetric landward water transport. The results are presented in Table 6.1. The percentage increase in return flows for low tide relative to high tide is presented in parentheses in the spring and mean low tide columns. These results indicate the relative effect of the tidal stage on return flow velocity (both high versus low and spring versus mean). They also indicate the importance of the relative depth over the bar (highest return flow velocities occurring for smallest clearance over the bar).

In interpreting the above results, it should be recalled that the calculations are based on the assumption of the waves breaking in water depth h_1 seaward of the bar crest and that the volume of water transported landward over the bar is a result of that breaking depth. Even if the waves break in depth h_1, the water transported landward over the bar may be less than that based on this assumption. To determine in part the effect of this consideration, the calculations were repeated with the assumption that the waves break on the bar crest and the water transported landward over the bar is associated with this breaking wave height. The relationship is presented in Equation (6.5) and the results in Table 6.2. Note that the return flow is greater at high tide than at low tide!

$$V_R = 0.08\sqrt{g(h_{Bar} + \eta)} \tag{6.5}$$

It is important to note that this seemingly inconsistent observation is somewhat an artifact of the depth-limited breaking assumption applied earlier. By using the simpler analysis in which water depth serves as a proxy for wave height, the effect of increased tidal stage and a fixed bar breakpoint is equivalent to assuming that the waves have grown larger; hence the higher tide case is by default associated with a larger shoreward mass transport which, although possible, is not necessarily true if offshore wave heights are limited. What is especially interesting about these results

TABLE 6.2

Bar Characteristics for Four Palm Beach Profiles (Figure 6.2) and Return Flow Velocities for Spring and Mean High and Low Tides

Profile Date	h_1 (m)	h_{Bar} (m)	V_R (m/sec) Spring Tide High	V_R (m/sec) Spring Tide Low	V_R (m/sec) Mean Tide High	V_R (m/sec) Mean Tide Low
1974	2.74	1.46	0.35	0.25 (−29%)	0.34	0.26 (−25%)
1990	4.27	2.13	0.40	0.32 (−20%)	0.40	0.3 (−17%)
2000	2.13	0.91	0.30	0.16 (−47%)	0.29	0.18 (−38%)
2006	2.74	1.52	0.35	0.26 (−28%)	0.35	0.27 (−24%)

Note: Wave breaking in water depth (h_{Bar} + tide level).

is that the same mechanism is also at play when assuming the waves are breaking at the offshore depth h_1. However the smaller percentage change in this large depth in comparison with the fixed tide range minimizes this effect in the Table 6.1 results and the percentage change in the smaller depth over the bar is the dominant effect, leading to much stronger return velocities at lower tide stages.

It is also worthwhile to revisit the criterion developed by O'Brien (1969) indicating that an approximate average tidal current through a sandy inlet of 1 ms^{-1} is required to maintain an inlet in an open and stable condition. It is reassuring that most of the velocities shown in Table 6.1 are of this magnitude, and interesting that none of the return velocities calculated in Table 6.2 reaches this magnitude. These simple results require further evaluation, yet they provide tentative insight into the mechanisms for rip channel initiation and the role of tide interaction with the bar. The simple model has shown that the conditions most conducive for rip channel initiation are for waves breaking seaward of the bar and for low tides and shallow clearance over the bar. Clearly, the same would be the case for maintenance of a rip current channel after it was formed.

RIP CURRENT CHANNEL MAINTENANCE

The preceding section focused on rip current channel initiation. After a channel is formed, the idealized current system changes from two-dimensional to three-dimensional; the seaward flow through the newly formed channel is now augmented with a portion of the landward water transport from adjacent barred areas, forming the well-known horizontal circulation cells shown in Figure 6.3. The other portion of the landward transport returns seaward over the bar due to the elevated water level inside the surf zone.

Thus, after initiation of a rip current channel, the required conditions to enlarge and maintain the channel are not as stringent as for its initiation. In fact, even with small waves, weak landward and seaward currents are most likely always present but may not be of sufficient magnitude to be considered rip currents. The rip current velocity for a case in which the waves break in water depth h_1 is given by:

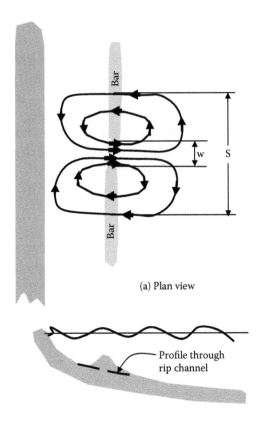

(a) Plan view

Profile through
rip channel

FIGURE 6.3   Landward flows over bar and concentrated seaward flows through a rip channel.

$$V_{RIP} = 0.08\alpha \left(\frac{S}{w} - 1 \right) \sqrt{g(h_1 + \eta)} \tag{6.6}$$

where α represents the portion of the landward transport over the bar that exits seaward through the rip channel and it is considered that the channel depth is equal to the depth h_1 at the location of the bar crest if the bar were not present. For purposes here, α is considered the same at high and low tides; but in actuality α would be somewhat greater at low tide. We do not attempt to quantify α here. If the waves break on the bar crest, the rip current velocity is:

$$V_{RIP} = 0.08\alpha \left(\frac{S}{w} - 1 \right) \left(\frac{h_{Bar} + \eta}{h_1 + \eta} \right) \sqrt{g(h_{Bar} + \eta)} \tag{6.7}$$

Table 6.3 shows the ratios of rip current velocities through the channel for spring and mean high and low tides for waves breaking at depth h_1 on the bar for the four Palm Beach profiles.

In summary, this section has shown that when a bar is present, waves breaking seaward of a bar crest are much more conducive to formation and maintenance of a rip

(Figure 6.2)

TABLE 6.3

Bar Characteristics for Four Palm Beach Profiles (Figure 6.2) and Ratios of Rip Current Velocities for Spring and Mean High and Low Tides

Profile Date	h_1 (m)	h_{Bar} (m)	Ratio of V_{Rip} for Wave Breaking at h_1 to Breaking at h_{Bar}			
			Spring Tide High	Spring Tide Low	Mean Tide High	Mean Tide Low
1974	2.74	1.46	2.14	3.50	2.19	3.28
1990	4.27	2.13	2.46	3.48	2.51	3.35
2000	2.13	0.91	2.57	7.52	2.68	6.25
2006	2.74	1.52	2.04	3.20	2.09	3.02

Note: Waves breaking at depth h_1 to waves breaking at h_{Bar}.

channel and generation of stronger rip currents. Rip current flows associated with the larger offshore breaking waves are clearly augmented by lower tidal stages but especially by the smaller bar depths (see year 2000 data in Table 6.3). This suggests that the ratio of bar height to depth in the absence of a bar may be a significant parameter relating to rip current occurrence, especially where waves break seaward of the bar.

STABILITY OF WADERS IN SURF ZONES

GENERAL NOTES

Results in this section and in Appendices A and B are presented in customary English units, in part because of the familiarity with pounds rather than Newtons as measures of weight. It is intuitive that a wader of 6 ft (1.8 m) height can maintain stability in greater water depths and in a particular current than one of, say, 3 ft (0.9 m) height, but we are unaware of any treatment of stability conditions that could be significant in educating swimmers of potential surf zone hazards.

Figure 6.4 illustrates the two types of stabilities considered. The first requires that the bottom friction experienced by the person be equal to or greater than the horizontal (hydraulic) drag force F_D of the current; the second is that the overturning moment induced by the drag force be less than the restoring moment by the individual. Appendices A and B apply simple hydrodynamic relationships of the velocities due to steady currents and oscillatory waves in the surf zone, with an idealized wader characterized by a circular cylinder of diameter D and length, ℓ as shown in Figure 6.4.

WADER STABILITY AGAINST CURRENTS

Appendix A examines the stability of waders against seaward-directed currents in the surf zone without the presence of waves. The first wader is 6 ft tall with a "diameter" of 0.8 ft; the second is 3 ft tall and has a "diameter" of 0.6 ft. Results are

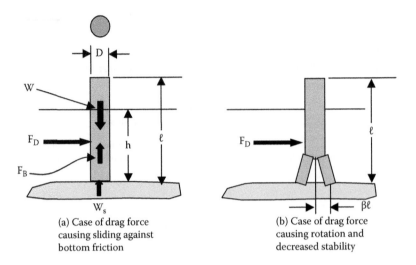

(a) Case of drag force
causing sliding against
bottom friction

(b) Case of drag force
causing rotation and
decreased stability

FIGURE 6.4   Two types of instabilities: (a) against sliding; (b) due to body rotation, resulting in increased buoyancy.

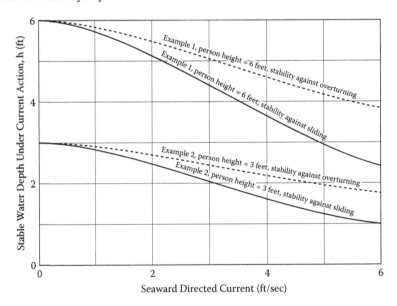

FIGURE 6.5   Two examples of limiting water depth for wader stability against seaward flowing currents.

presented in Figure 6.5. For the assumptions described in Appendix A, a wader of 6-ft height would be stable against sliding for a current of 4 ft/sec or less in a water depth of 3.7 feet or less. For the same wader and water depth, stability against overturning would occur for a current exceeding 6 ft/sec. A 3-ft wader would be stable against sliding for a current of 4 ft/sec or less in a depth of 1.6 ft or less and for

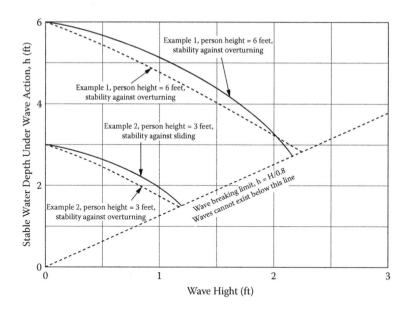

FIGURE 6.6 Two examples of limiting water depth for wader stability against waves and wave-breaking criterion in which $\kappa = 0.8$. Results shown for wave crest are less stable than those shown under the trough for a wader.

the same water depth, stable against overturning in a current greater than 6 ft/sec. Considering that the velocity threshold for rip currents is approximately 3 ft/sec, the rule of thumb that an inexperienced swimmer should not venture out into the surf zone beyond half his height appears reasonable.

WADER STABILITY AGAINST WAVE-INDUCED WATER PARTICLE VELOCITIES

Appendix B also investigates wader stability, but the destabilizing cause is the drag force resulting from wave-induced oscillatory water particle velocities. As for Appendix A, seconds values are based on simple wave theory and illustrated with examples of wader heights of 3 and 6 ft and the diameters noted. These results are presented in Figure 6.6. A wader of 6-ft height would be stable against sliding in a water depth of 4 ft for all wave heights less than approximately 1.7 ft and stable against overturning at a depth of 4 ft for all wave heights below 1.5 ft. A wader of 3-ft height would be stable against sliding in a water depth of 2.5 ft for all wave heights less than about 0.6 ft and against overturning for all wave heights less than about 0.4 feet. As shown in the figure, the wave heights are limited by water depth such that they cannot exist below the straight dotted line.

IMPROVING DETECTION AND PREDICTION OF RIP CURRENTS

As discussed previously, rip current predictions provide valuable warning services. However, these predictive indices used lifeguard-based rip rescue records as proxy data for the actual incidence of rip currents in order to develop a sufficiently large database for index calibration. A host of difficulties surround this proxy representation and hence it would be useful if some rapid method could be developed to actually determine whether rip currents are present on a specific day and at a particular location.

Better Detection of Rip Currents through Thermal Remote Sensing

The water temperatures in rip currents may be slightly different from those of water outside the surf zone, possibly providing a basis for detection by airborne thermal sensors. It is our understanding that this method has been successfully applied in Japan. GPS coupled with thermal imaging on an airborne platform would allow the locations of the rip currents to be established daily or more often if desired. Additionally, if this method were found effective, the thermal imagery might be applied to determine the strengths of the rip currents identified. At a minimum, development of a database of rip currents in this manner would call attention to areas especially prone to rip currents. Finally, correlation with physical parameters (waves, tides, etc.) should lead to improved understanding of the genesis of rip currents in general and improvement of prediction methods in particular.

Need for Including Spatial Information in Rip Current Prediction

As noted, existing rip current predictions are useful, but often the same prediction applies over a very large region (in some cases, >100 km). Within the prediction region, local features may create areas that are more or less prone to rip current occurrence than average locations. Predictions issued over wide areas may not be valued due to the perception that they represent "overwarnings" or lack of location specificity. Thus, it is recommended that an effort be made to include more detailed spatial detail in future rip current indices. The thermal sensing approach discussed previously, if effective, could help provide the type of spatial detail desired.

UP-TO-DATE ADVISORIES

Once a family reaches the beach, attention turns to entering the water; caution related to safety and discussions of the possible presence of rip currents and appropriate responses may become secondary. We have all seen signs along highways indicating the availability of local traffic information on designated AM radio stations. In areas near concentrations of surf zone recreation, it should be possible to have such radio stations dedicated to swimming and rip current advisories available to families approaching the beach. In some areas these signs could be located on or near causeways to barrier islands where young swimmers are "captive" for 5 or 10 min prior to arriving at a beach.

In addition to information about the likelihood or presence of rip currents, the advisories could provide related information including locations of guarded beaches, areas suitable for experienced swimmers only, and techniques for escaping offshore-directed currents. Such information would be valuable to both experienced and less qualified swimmers and also inform chaperones of the hazards and needs to take precautions.

Additionally, this information would reinforce any previous rip current-related education and remain fresh in the minds of those likely to encounter rip currents.

PERSONAL RESPONSIBILITY AND EDUCATION

It is doubtful that all beaches will ever be completely protected by lifeguards. Some individuals prefer to swim in unguarded areas and/or to swim at night. It is possible that with more lifeguards present, individuals will consciously or subconsciously take less personal responsibility. Clearly, accepting greater personal responsibility will contribute to fewer drownings. Improved approaches to increase personal responsibility are presently limited to educating and training candidate swimmers in realistic conditions and educating the chaperones of young swimmers about the risks associated with and the appropriate responses to rip currents. One approach emphasized at the First International Rip Current Symposium was more and improved signage.

LEGAL DILEMMA

A goal of all beachfront hotels is the safety of their clienteles. Many beachfront hotels in Florida have erected signage and employ lifeguards. Unfortunately, from a legal perspective, a hotel taking substantial precautions relative to rip current hazards can be interpreted legally as accepting liability for the related safety of its clientele. Thus, from a purely legal standpoint, it is advisable not to cross this threshold regarding beach safety. Of course, this places greater responsibility on individuals and emphasizes the need for rip current warnings characterized by greater temporal and spatial specificity.

CONCLUSIONS

Several issues have been addressed relating to the role of proportion of depths occupied by bars, hydrodynamics of rip channel formation and maintenance, wader stability within the surf zone against steady currents and oscillating velocities due to waves, and other surf zone safety and legal points. It is hoped that the role of bar proportion will be investigated with the goal of determining whether its inclusion could contribute to more effective prediction models. The stability relationships were based on simple models and wader idealizations. These relationships should be investigated and refined with the intent to improve swimmer education and safety. Other issues discussed that have the potential of reducing surf zone hazards include improved rip current identification, more detailed prediction methods, education, and improved advisories.

REFERENCES

Dalrymple, R.A. 1978. Rip currents and their causes. *Proc. 16th Intl. Conf. on Coastal Engineering*. American Society of Civil Engineers. Hamburg, pp. 1414–1427.
Engle, J.A. 2003. Formulation of a Rip Current Forecasting Technique through Statistical Analysis of Rip Current-Related Rescues. M.S. thesis, University of Florida, Gainesville.
Lascody, R.L. 1998. East Central Florida rip current program. *Natl. Weather Dig.*, 22: 25–30.

Lushine, J.B. 1991. A study of rip current drownings and related weather factors. *Natl. Weather Dig.*, 16: 13–19.
O'Brien, M.P. 1969. Equilibrium flow areas of inlets on sandy coasts. *J. Waterways Harbors Div.*, 95: 43–52.

APPENDIX A: STABILITY OF WADERS IN SEAWARD-DIRECTED STEADY CURRENTS

INTRODUCTION

The interaction of waders with surf zone currents was investigated with a goal of establishing the limiting water depths in which individuals can maintain stability in the presence of seaward-directed steady currents. Methods are highly simplified and approximate but provide results that may be useful.

IDEALIZED WADER CONCEPT

To accomplish the goal set for this investigation, it was necessary to idealize a wader that could be readily treated hydrodynamically. For this purpose, we consider the wader as a circular cylindrical shape with height ℓ, diameter D as shown in Figure 6.4a, and uniform mass density ρ_B. The methodology may be applied to more realistic body characterizations including an actual body planform (which is more elliptical than circular) and variations in body dimensions and density from the feet to the head. However, for our purposes here, this idealization will allow presentation of concepts without the burden of complicating details. For the idealized considerations, the diameter of the individual for a body weight W is:

$$D = \sqrt{\frac{4W}{\rho_B g \pi \ell}} \qquad (A.1)$$

SUBMERGED WEIGHT IN SURF ZONE

The submerged weight W_s of an individual at a depth h in the surf zone is:

$$W_s = \rho_w g \frac{\pi D^2 h}{4}\left(\frac{\rho_B \ell}{\rho_w h} - 1\right) \qquad (A.2)$$

Note that the individual would float ($W_s = 0$) when the left side of Equation A.2 is zero and the terms in the parentheses balance.

CURRENT FORCES ON INDIVIDUAL

The current forces are represented by the so-called drag force calculated as:

$$F_D = \frac{C_D \rho_w D}{2} h V^2 \qquad (A.3)$$

LIMITING WATER DEPTH FOR STABILITY

Figure 6.4 illustrates the two types of stabilities against currents that may be examined. For the case of *stability against sliding*, the bottom friction is equal to the product of the submerged weight W_s and the friction coefficient μ, resulting in:

$$F_D = \mu W_s$$

$$\frac{C_D \rho_w D}{2} h V^2 = \mu \rho_w g \frac{\pi D^2 h}{4} \left(\frac{\rho_B \ell}{\rho_w h} - 1 \right) \tag{A.4}$$

from which the stable depth h can be expressed as:

$$h = \frac{\rho_B}{\rho_w} \ell \left(\frac{1}{1 + \dfrac{2C_D}{\pi \mu g D} V^2} \right) \tag{A.5}$$

For the case of *stability against rotation*, the restoring moment must be equal to the overturning moment due to the drag force considered to act at mid-water depth. This balance can be expressed as:

$$F_D \frac{h}{2} = \beta \ell W_s$$

$$\frac{C_D \rho_w D}{4} h^2 V^2 = \beta \ell \rho_w g \frac{\pi D^2 h}{4} \left(\frac{\rho_B \ell}{\rho_w h} - 1 \right) \tag{A.6}$$

in which β represents the proportion of wader height that acts as a lever arm with the wader submerged weight to resist the overturning moment of the drag force (see Figure 6.4b). This equation is quadratic in h with the solution:

$$h = 0.5 \left(-B + \sqrt{B^2 + 4B \frac{\rho_B}{\rho_w} \ell} \right) \tag{A.7}$$

in which $B = \dfrac{\pi g D \beta \ell}{C_D V^2}$.

CHARACTERISTICS CONSIDERED

These equations were evaluated for two body dimension examples: $\ell = 6$ ft and 3 ft, considering body diameter to vary approximately as $D \propto \ell^{1/2}$ such that the diameters D were 0.8 and 0.6, respectively. The following conditions are considered in the examples: $C_D = 1.0$, $\mu = 0.6$, $\rho_B/\rho_w = 1.0$, $\beta = 0.5$. Results are presented in Figure 6.5 and discussed in the main body of this text.

APPENDIX B: STABILITY OF WADERS IN OSCILLATORY WAVE CURRENTS

Waves induce oscillatory water particle velocities with shoreward-directed velocities under the wave crests and seaward velocities under the wave troughs. It is of interest to compare the stabilities of waders in various depths of water when affected by water wave particle velocities. Again, linear shallow water wave theory will be used along with the same wader idealization. The oscillatory velocity under a wave of height H is:

$$V_W = \pm \frac{H}{2} \sqrt{\frac{g}{h}} \tag{B.1}$$

where the plus sign applies under the wave crest and the minus sign under the wave trough. The horizontal drag forces are:

$$F_D = \pm \frac{C_D \rho_w g D H^2}{8} \frac{(h \pm H/2)}{h} \tag{B.2}$$

The immersed weight W_s in the presence of waves is:

$$W_s = \rho_w g \frac{\pi D^2}{4} \left(\frac{\rho_B \ell}{\rho_w} - (h \pm H/2) \right) \tag{B.3}$$

The relationship for stability against *sliding* is:

$$h_S = \frac{-B_S + \sqrt{B_S^2 - 4C_S}}{2} \tag{B.4}$$

in which

$$B_S = \frac{\left(1 + \delta_S \dfrac{H}{2} - \delta_S \dfrac{\rho_B}{\rho_w} \ell \right)}{\delta_S} \tag{B.5}$$

and

$$C_S = \frac{H}{2\delta_S} \tag{B.6}$$

where

$$\delta_S = \frac{2\pi D \mu}{C_D H^2}$$

(B.7)

For stability against *overturning*, the limiting water depth is

$$h_{OT} = \frac{-B_{OT} + \sqrt{B_{OT}^2 - 4C_{OT}}}{2}$$

in which:

$$B_{OT} = \frac{\left(H(1+\delta_{OT}) - \delta_{OT}\frac{\rho_B}{\rho_w}\ell \right)}{1+\delta_{OT}}$$

(B.8)

and

$$C_{OT} = \frac{H^2}{4(1+\delta_{OT})}$$

(B.9)

where

$$\delta_{OT} = \frac{4\pi D \beta \ell}{C_D H^2}$$

(B.10)

We consider the same two examples as in Appendix A for body characteristics and μ and β values. Graphic results are presented in Figure 6.6 and discussed in the main text for the wave crest phase position because the wader would be less stable under the crest due to greater immersion (and thus reduced submerged body weight), increased depth over which the drag forces act, and increased level arm of the drag force.

7 Florida Rip Current Deaths
Forecasts and Statistics

James B. Lushine

CONTENTS

INTRODUCTION

Official statistics from the National Weather Service (NWS) show that in the past 20 years the state of Florida has had more deaths from rip currents than from more notorious natural phenomena such as hurricanes, lightning, and tornadoes combined (Figure 7.1). Various estimates of the number of U.S. rip fatalities range from 35 to 200 annually. Worldwide, the number of rip current-related deaths is not well known, but an estimated 155 nations face rip currents.

Lushine (1991) initially examined medical examiner death records in southeast Florida for likely occurrences of rip drownings in Miami-Dade, Broward, and Palm Beach counties from 1979 to 1988. Using data from that period, an empirical relationship between rip current-related deaths and certain weather and oceanographic conditions led to the creation of the Lushine Rip Current Scale (LURCS) used to predict the daily risks of dangerous rip currents along the southeast Florida surf beaches. Lascody (1998) developed a technique for assessing rip risks along the east central Florida coast; further north, the Mid-Atlantic LURCS (MALURCS) was used to express the danger. NWS offices elsewhere in Florida and in other states and U.S. possessions have also conducted such studies, culminating in implementation of a uniform nationwide NWS policy to issue seasonal daily surf forecasts that include the risks of rip currents. These surf forecasts may be accessed by the general public, local beach patrols, and other beach safety organizations through web sites and are broadcast on NOAA weather radio.

RIP CURRENT DEATHS

In southeast Florida, rip current-related drowning statistics from 1979 through 2008 were compiled using death certificates issued by medical examiner offices for

125

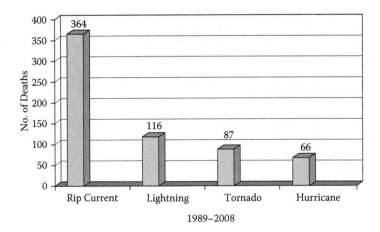

FIGURE 7.1 Comparison of Florida rip current deaths with other state weather hazards. Rip current deaths (364) exceed the number of deaths (269) from lightning, tornados and hurricanes (including other tropical cyclones) combined.

Miami-Dade, Broward, and Palm Beach Counties (Lushine (1991). Beginning in 1994, rip current deaths and injuries were included in NOAA *Storm Data* reports for the state. NWS meteorologists compile local death, injury, and damage statistics for *Storm Data* reports. Table 7.1 lists the sources covering rip-related death data cited in this chapter.

Storm Data entries are largely based on media input; therefore, newsworthy events, such as hurricanes, tornadoes, and river floods, are likely to be included. By contrast, deaths from such phenomena as lightning and rip currents that affect few people are less likely to be reported. Also rip current and lightning deaths often occur over widely separated geographic areas, making the likelihood of media coverage

TABLE 7.1
Summary of Sources of Data on Rip Current–Related Deaths

Location	Source	Dates
Southeast Florida	M.E. Office	1979 through 1988
State of Florida	Newspaper	1989 through 1994
State of Florida	*Storm Data*	1994 through 2008
United States	*Storm Data*	1999 through 2008

Notes: M.E. indicates medical examiner records in Miami-Dade, Broward, and Palm Beach Counties. *Storm Data* is a monthly NOAA publication.

less reliable. For example, Lushine (1996) compared Florida lightning deaths cited in NOAA's *Storm Data* publication and the Florida Office of Vital Statistics (FOVS) records and found 31% under-reporting in *Storm Data*. Inputs to FOVS records for lightning are frequently based on the International Classification of Diseases (ICD) that maintains a specific classification for lightning deaths. Similarly, Ashley (2009) found 29% under-reporting for lightning deaths in *Storm Data* for the United States. A similar comparison of FOVS and *Storm Data* for rip death statistics cannot be made because the ICD does not have an explicit classification for rip currents as a cause of death. A conservative estimate is that *Storm Data* underestimates the number of true rip current-related deaths by at least 50% and perhaps as much as 100%.

TIDAL INFLUENCE

LURCS and similar techniques take into account the fact that rip current deaths are more prevalent around the times of daily low tides. Brander and Short (1999) showed that rip speeds are greater near low tide times than at other daily tidal times. A faster rip current is potentially a more deadly one. Komar (1976) also suggested that more deaths may occur near the times of low tides because bathers and poor swimmers may be tempted to venture further offshore then and thus be more susceptible to rips. Figure 7.2 shows the relationship between rip deaths and daily tidal levels in southeast Florida from 1979 through 2008. Fifty percent of the rip current deaths occurred within 1 hr of low tide.

The southeast Florida coast has a semi-diurnal tidal cycle with usually two low tides per day. The original LURCS calculation did not distinguish which of the daily low tides was occurring. Using data from 1994 through 2008, the tidal calculation was segregated into the daily lower low tide and the daily higher low tide. About twice as many rip current-related deaths occurred near the daily lower low tide compared to the daily higher low tide (Figure 7.3).

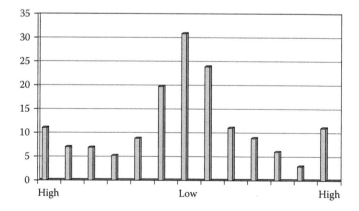

FIGURE 7.2   Rip current deaths related to tidal times from 1979 to 2008.

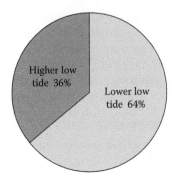

FIGURE 7.3   Percentage of rip current deaths related to daily low tide.

DROWNING STATISTICS

Florida has the most rip current deaths of any state or possession in the United States, with an estimated 162 victims from 1999 through 2008 (Figure 7.4). California was second with 41 deaths while Hawaii (38), Guam (34), and North Carolina (27) round out the top five. NWS offices in the Hawaiian Islands do not categorize rip drownings specifically in *Storm Data*. Instead, they use the more generic "high surf"

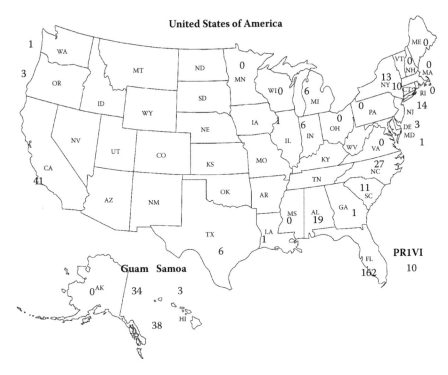

FIGURE 7.4   Rip current deaths by state and possession reported by *Storm Data* from 1999 to 2008.

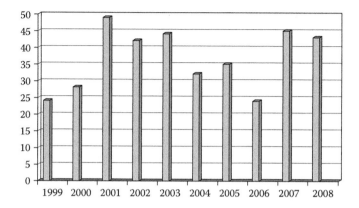

FIGURE 7.5   Estimated annual number of rip current deaths in the United States and its possessions from *Storm Data* records.

classification to estimate rip deaths. The annual number of rip current-related deaths is not actually known. Estimates from *Storm Data* are shown in Figure 7.5. Other estimates include the following:

- Lushine (1991) used indirect statistics to estimate that rip deaths from 1979 through 1988 were 100 to 200 annually in the U.S.
- The United States Lifesaving Association (USLA) estimates annual rip current deaths at 100 per year.
- Gensini and Ashley (2009) used Centers for Disease Control and Prevention data to calculate the number of U.S. rip current deaths annually at 35.

Southeast Florida often exhibits large year-to-year variations in rip drownings. Figure 7.6 shows the estimated deaths from 1979 to 2008. The total number of deaths during the 30-year period was 219, with a maximum of 22 and a minimum of 0. A linear regression (trend line) computed for southeast Florida rip current deaths showed a decrease from 11 in 1979 to 4 in 2008. Figure 7.6 also shows the trend

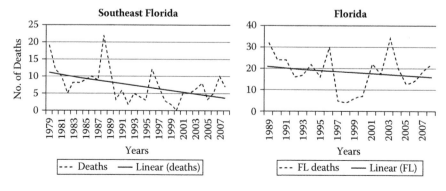

FIGURE 7.6   Trends in rip current deaths (left) in southeast Florida from 1979 to 2008 and (right) entire state from 1989 to 2008.

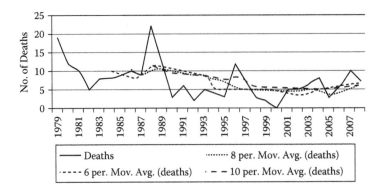

FIGURE 7.7   Running mean of rip current deaths in southeast Florida from 1979 to 2008.

line for the state, indicating a decrease in rip current deaths from 21 in 1989 to 16 in 2008. A 4-year running mean of rip current deaths in southeast Florida from 1979 through 2008 was plotted (Figure 7.7). This graph indicates an approximate 10-year periodicity of rip current deaths.

CONCLUSIONS AND RECOMMENDATIONS

The annual number of rip current-related deaths in the United States during the past decade is estimated at 60 to 100, but many limitations surround the data gathering. While the trend in Florida is downward, large year-to-year variations in rip drowning incidence occur. It is speculated that heightened awareness by beachgoers because of greater publicity from the media and surf forecasts by the NWS have helped to reduce the death toll. Another initiative to make the public more aware of the rip current danger has been implementation of a uniform beach flag system in Florida.

A 10-year cycle of rip current deaths in southeast Florida may be related to the sunspot cycle. Meehl et al. (2009) showed that winds over tropical areas are stronger near the maximum time of the 11-year sunspot cycle. An increase in wind speed could account for increases in deadly rip currents. It is recommended that:

- LURCS calculations should be adjusted to account for the greater possibility of rip current deaths at daily lower low tides.
- NWS forecast offices should incorporate Office of Vital Statistics information on rip deaths into *Storm Data* reports.
- A specific code for rip current deaths should be incorporated into the International Classification of Diseases.

ACKNOWLEDGMENTS

The author thanks Rusty Pfost and Robert Molleda of the National Weather Service Forecast Office in Miami for creating graphics and providing access to rip current data.

REFERENCES

Ashley, W.S. and C.C. Gilson. 2009. A reassessment of U.S. lightning mortality. *Bull. Amer. Meteor. Soc.,* 90: 1501–1518.

Brander, R.W. and A.D. Short. 1999. Morphodynamics of a large-scale rip current system, Muriwai Beach, New Zealand. *Marine Geol.,* 165: 27–39.

Gensini, V.A. and W.S. Ashley. 2009. An examination of rip current fatalities in the United States. www.springerlink.com/index/16T26655t504232.pdf. Published online Sept. 23, 2009.

Komar, P.D. 1976. *Beach Processes and Sedimentation.* Prentice-Hall, New York.

Lascody, R.L. 1998. East Central Florida rip current program. *Natl. Weather Dig.,* 22: 25–30.

Lushine, J.B. 1991. A study of rip current drownings and related weather factors. *Natl. Weather Dig.,* 16: 13–19.

Lushine, J.B. 1996. Under-reporting of lightning deaths in Florida. *Intl. Lightning Detection Conf., Tucson, Arizona,* Global Atmospherics.

Lushine, J.B. 2000. Under-reporting of heat and cold deaths in Florida. 25th National Weather Association Annual Meeting, Gaithersburg, MD.

Lushine, J.B. 2005. A blueprint for reducing rip current deaths in the United States. *Solutions to Coastal Disasters,* American Society of Coastal Engineers, pp. 257–263.

Meehl, G.A., J.M. Arblaster, K. Matthes et al. 2009. Amplifying the Pacific climate system response to a small 11-year solar cycle forcing. *Science,* 325: 1114–1118.

8 Remote Sensing Applied to Rip Current Forecasts and Identification

Brian K. Haus

CONTENTS

INTRODUCTION

The use of remote sensing to enhance public safety is well established for many types of natural hazards. Satellites and aircraft monitor hurricane development and Doppler weather radars are the key components of tornado detection and warning. Implementation of remote sensing technologies to improve beach safety is not as well developed. Remote observations of rip currents or waves and currents that are most likely to generate rip currents offer the promise of improving beachgoer safety.

In situ instrumentation such as current meters and wave buoys provide useful information, but several complicating factors must be considered. The primary limitation is that dense arrays of observation points may be required to provide the necessary spatial coverage for even a moderate-sized beach. Also, installing and maintaining in situ observation equipment in a surf zone is a difficult and expensive undertaking.

Remote sensing technologies offer the promise of much larger coverage areas and simpler implementation in coastal areas. These technologies may provide improved initial, boundary, and forcing conditions for models and direct observation of rips when they occur. Which approach is most useful at a particular beach depends on local oceanographic conditions, beach and nearshore topography, and stakeholder resources and requirements. This chapter will explore the application of promising remote sensing techniques for improving rip current observation and forecasting. The primary focus will be on land-based high frequency (HF) radars and microwave radars because of their comparatively low cost and capability for continuous monitoring.

BACKGROUND

Two different classes of rip currents need to be considered for remote sensing applications. Some rip currents are fixed in location by a hardened topographic feature, whether man-made (jetties, seawalls, etc.) or natural (rock outcrops, pocket beaches, reefs, etc.). Others vary in location, particularly on open sandy coasts. Time scales of rip variability due to incident waves range from minutes (MacMahan et al., 2004a; Reniers et al., 2006) when forced by infragravity waves to very low frequency, wave-driven oscillations at scales greater than a few minutes (Reniers et al., 2007; MacMahan et al., 2004b).

Longer term changes in rip currents occur on hourly scales due to tidal changes in water level relative to nearshore topography (Brander, 1999; Brander and Short, 2001; MacMahan et al., 2006). At even longer temporal scales, wave conditions vary with regional winds due to synoptic systems (typically 3 to 5 days in mid-latitudes) and seasonal changes. Beach morphological changes due to erosion, accretion, or redistribution of sand can also modify local rip currents (Brander and Short, 2001).

The two modes of transfer of information about rip currents are forecasting and identification. Forecasting relies on knowledge (historical, statistical, or dynamical information from models) of a coastal system to predict when rip currents will occur at a particular location. As forecasts of any dynamical phenomenon are inherently uncertain, a probability of occurrence should also be attached; it is critical to strike an appropriate balance between safety and inconvenience in the United States. The National Weather Service (NWS) offices make surf forecasts, but rip variability relative to local winds and waves poses some challenges for developing accurate forecasts.

The second mode of information is identification— directly observing a rip current in a particular location. Recently an on-duty lifeguard was observed warning swimmers to stay out of an area where a rip current was clearly visible. The heavy holiday weekend crowd of beachgoers was temporarily inconvenienced, but possible loss of life was avoided. The application of remote sensing technologies to this realm of rip current warnings seeks to extend the trained eyes of beach safety professionals beyond the visible stretch of water in front of their lifeguard towers. The challenge then becomes to transfer the information in a timely fashion and reduce the rate of false positives so that lifesavers and beachgoers do not lose confidence in the system.

RIP CURRENT FORECASTING

Two key components of developing an accurate rip current forecast for a particular beach are the incident waves and nearshore topography. In the case of fixed rips, the topography is paramount to appropriately model tidal elevation changes. For regional-scale forecasts of rip current hazards, historical topographic information may be sufficient. In these situations, forecasting becomes a two-fold challenge: forecast the incident deepwater wave conditions and translate this information into a probability of rip current occurrence.

INCIDENT WAVE CONDITIONS

High wave energy conditions are related to the strongest rip currents and constitute the most elevated hazards for beachgoers. This basic premise underlies most applications of rip forecasts by the National Weather Service (NWS). While the shallow water wave and current models available for routine rip current forecasts are limited, remote sensing can be implemented to provide improved forecasts. The main challenges are to: (1) provide a reliable deepwater wave field, (2) incorporate this information into boundary conditions for high resolution coastal models, and (3) force the model with accurate regional wind fields. Remote sensing technologies can be used to help meet each of these challenges in particular circumstances.

Basin-scale wave forecasts at a resolution of 0.25 degrees (Figure 8.1a) using the Wavewatch III model (Tolman et al., 2002) are available throughout the U.S. (http://polar.ncep.noaa.gov/waves/main_int.html). Ocean circulation forecasts are also available, such as the forecast system using a hybrid coordinate model (Mehra and Riven, 2010). In regions with relatively simple topography and weak currents, wave and current conditions can be used directly as inputs to wave models for hindcasts and forecasts. Where the shoreline is complex or offshore currents are strong, deepwater wave values may not be appropriate for direct use in nearshore wave models. The southeast Florida coast from Cape Canaveral to the Dry Tortugas is such an area because the Florida Current significantly modifies the offshore wave climate (Haus, 2007).

High frequency Doppler radars operate in frequency ranges of 5 to 50 MHz, propagate along the air–sea interface, and preferentially scatter off ocean surface waves with lengths between 3 and 30 m (Figure 8.1b). They offer the promise of providing incident deepwater wave conditions over regions of the coastal ocean where in situ observations are unreliable or absent. For rip forecasting, it is essential that full frequency-directional wave spectra be observed; simple wave parameters or frequency spectra alone do not provide sufficient information in most circumstances.

Barrick (1972) derived expressions that relate first- and second-order reflected radar signals to the ocean surface wave energy spectra. Methods to invert the Barrick equations and retrieve the surface wave directional energy spectra were developed by Lipa (1978), Lipa and Barrick (1986), Wyatt (1990), Howell and Walsh (1993), and Hisaki (1996). Wyatt et al. (1999) demonstrated that when using two radar sites, inversion methods applied to phased-array HF radar systems can yield wave directional spectral estimates comparable to those recorded by moored buoys. Directional spectra observed in this manner (Figures 8.1c and 8.1d) can then be used as inputs to coastal wave models such as SWAN (Voulgaris et al., 2008).

Many problems are associated with installation and maintenance of in situ wave measurements in strong currents (Voulgaris et al., 2008). The capability of Wellen radar (WERA) phased-array systems to remotely provide simultaneous current (Shay et al., 2007) and wave observations (Haus et al., 2010) is very attractive. Such an approach can provide larger spatial coverage than is feasible using point measurement

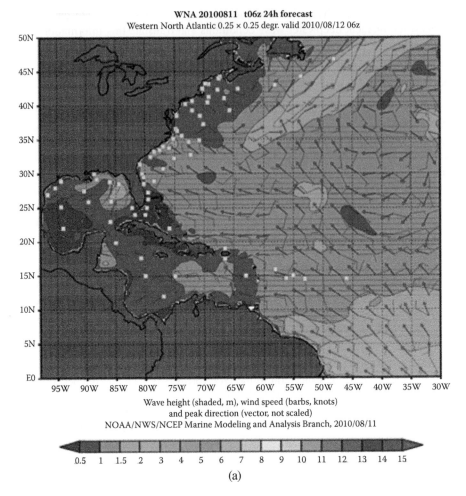

WNA 20100811 t06z 24h forecast
Western North Atlantic 0.25 × 0.25 degr. valid 2010/08/12 06z

Wave height (shaded, m), wind speed (barbs, knots)
and peak direction (vector, not scaled)
NOAA/NWS/NCEP Marine Modeling and Analysis Branch, 2010/08/11

0.5 1 1.5 2 3 4 5 6 7 8 9 10 11 12 13 14 15

(a)

FIGURE 8.1   *(See color insert.)* (a) Example of Western North Atlantic wave hindcast provided by NOAA/NWS/NCEP. (b) Principle of operation of HF radar: transmitted radar energy propagates over the air–water interface and is reflected back to the transmitter preferentially by waves of half the radar wavelength. (c) Sample wave spectrum derived from the University of Miami HF radar system using Seaview sensing software. (d) Modeled spectrum by SWAN utilizing local wind and current fields (from HF radar) as initial conditions. (Courtesy of Werner Gurgel, University of Hamburg.)

systems (Figure 8.2a). In many cases HF radars are positioned to optimize surface current coverage at the expense of a limited region of appropriate overlap for dual-site wave retrievals (Wyatt et al., 2005) or are not set to sample optimally for wave spectral measurements (Wyatt et al., 2009). Nonetheless, with appropriate consideration and modification of deployment and sampling strategies to mitigate these complications, they can fill gaps inherent with in situ data networks. Furthermore, land-based systems (Figure 8.2b) significantly reduce the operational difficulties related to data transfer, processing, and archiving.

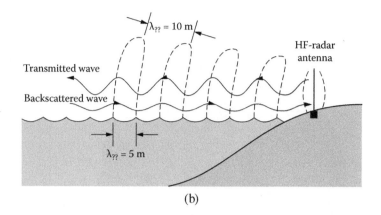

(b)

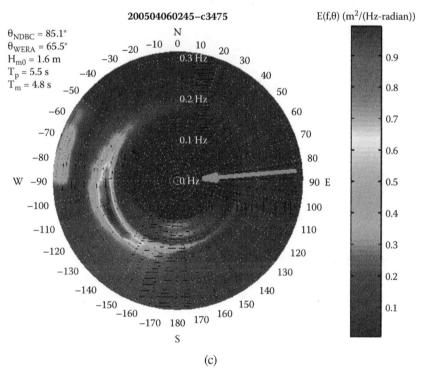

(c)

FIGURE 8.1 (continued).

It is imperative that prior to adopting these methods for operational purposes, evaluation for reliability and accuracy be undertaken as discussed in the NOAA–NDBC
National Waves Plan (NOAA, 2009). This plan calls for research into applications
of HF radar technology to supplement gaps in coastal wave observations that will
provide guidance for the appropriate filtering and error structures of the input wave
height and current fields necessary for inclusion in the Integrated Ocean Observing

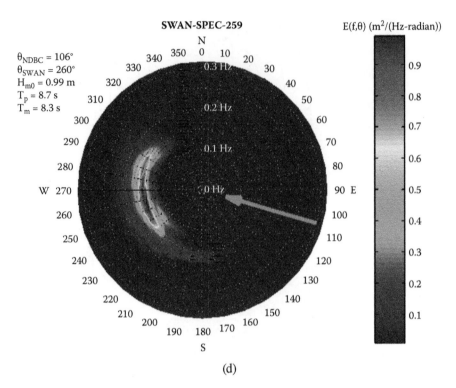

(d)

FIGURE 8.1   (continued).

System (IOOS) framework. Improved local wave modeling can then be used to improve understanding and forecasts of rip currents (Reniers et al., 2006, 2007).

REGIONAL TOPOGRAPHY, LARGE-SCALE CURRENTS, AND WIND FIELDS

Given a reliable deep-water wave field and depending upon local circumstances, three other key types of information may be required to derive accurate local rip current forecasts: (1) sub-areal and exposed topography of the beach area, (2) coastal wind field (particularly in mountainous regions), and (3) mean current field, especially near inlets and river mouths. Remote sensing technologies can be useful for defining currents that directly affect rip currents close to shore or exert indirect effects through wave–current interactions.

Topography

The options for routinely deriving sub-surface topography from remote platforms unfortunately are much more limited than the choices available for analyzing winds, waves, and currents. This is primarily because most electro-magnetic signals do not penetrate water and can observe only surface properties. This is also true for passive optical systems that have been widely exploited to infer topography in surf zone applications (Holman and Stanley, 2007). Identifying the locations of wave breaking (Lippman and Holman, 1989) or changes in the local wave field (Splinter

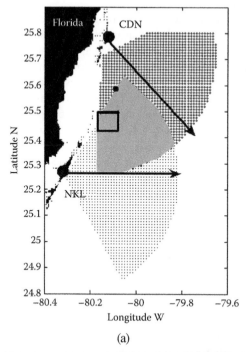

(a)

(b)

FIGURE 8.2   (a) Current and wave coverage areas for WERA HF radar deployment in the Straits of Florida (*Source:* Voulgaris, G., B.K. Haus, P. Work et al. 2008. *Marine Technol. Soc. J.,* 42: 68–80. With permission.) (b) WERA HF radar receive antenna array installed at Crandon Beach, Miami-Dade County, Florida.

and Holman, 2009) can be used to infer the locations and depths of sub-surface bars, respectively (Figure 8.3).

Airborne LIDAR (light detection and ranging) systems are the most robust, accurate, high resolution technologies available for coastal topographic mapping (Robertson et al., 2004; Irish et al., 2000). They are routinely used to map shoreline position changes and topography over coastal regions in the U.S. and elsewhere and are particularly useful for post-event surveys to assess coastal changes after storms and tsunamis (Gibeaut, 2003). Specialized LIDAR systems are able to achieve penetration into the water column and derive sub-surface topography at vertical resolutions on the order of 10 to 15 cm to depths ~10 m. However, LIDAR signals are rapidly attenuated by both particles and turbulence in the water column (Churnside, 2008). The main limiting factor for routine applications of LIDAR for topographic mapping where coastal conditions change rapidly is the high cost of flying repeated airborne surveys and processing large volumes of data.

Wind Fields

Scatterometers measure wind fields by observing the backscattered energy from the air–sea interface at a wavelength range from ~1 to 30 cm. Calibrated backscatter can be related to wind speed through a geophysical model function (GMF). In the open ocean, large (kilometer scale) images mitigate the backscatter caused by various processes such as surface currents, surfactants, and long waves. In the coastal ocean, the need for more spatially resolved wind fields makes these issues more pressing. Scatterometer-measured wind vectors have been assimilated into wind field data that are now widely available for the entire globe (Bentamy et al., 2002). The main complication for surf zone application is that satellite-mounted scatterometers cannot sample within ~25 km of the coast; consequently they cannot resolve small-scale coastal wind variability. Aircraft-mounted radars can achieve high spatial resolution, but are typically limited to dedicated field campaigns (Plant et al., 2005).

Large-Scale Coastal Circulation

The same high frequency (HF) radar systems that measure local wave fields can also be used to define the larger scale circulation features in a particular region. HF radar experiments have revealed many complex surface currents that were little understood or not previously observed through the use of traditional single-point measurements. Examples include tidal residual eddies that form near the mouth of the Chesapeake Bay (Marmorino et al., 1999) and small-scale vortices formed during the relaxation of a coastal buoyancy current along the Outer Banks of North Carolina (Haus et al., 2003). These velocity structures may have many different length scales and translation velocities. Recently HF radar has been broadly implemented in ocean observing systems (Shay et al., 2007 and 2008), and for routine observations of surface currents along the U.S. coastlines.

RIP CURRENT IDENTIFICATION

The two main technical challenges that must be overcome when attempting direct observation of rip currents using remote systems are latency and resolution. The

(a)

(b)

FIGURE 8.3 Surf zone images from Argus system deployed at U.S. Army Corps of Engineers Field Research Facility at Duck, North Carolina. (a) Snapshot taken toward the south from the facility tower on August 11, 2010 at 11:00 EST. Offshore significant wave height was ~2 m. (b) Ten-minute average of snapshots. Note persistent cross-shore features.

spatial resolution of satellite-based remote sensing has steadily increased to the point where it can be applied to rip current studies. Synthetic aperture radar missions such as TerraSarX have achieved resolutions as fine as 2 m, but relatively long repeat cycles and large amounts of time required to transfer and process data will continue to be limiting factors. Ground-based microwave systems are not limited by repeat cycles or data transfer because they may be distributed directly to local computer networks. Consequently, this technology is very promising for rip identification if resolution and sampling domain issues can be resolved.

SATELLITE SYSTEMS

Optical remote sensing observations from satellites have revealed patterns in near-shore regions that are strongly indicative of rip currents. For example, detections of local regions of turbid water extending offshore of the surf zone are similar to those observed by local optical systems (Figure 8.3a). The high resolution of optical remote sensing (sub-10 m) would enable these qualitative observations to be of considerable use for rip identification but the time between successive over-flights of the same location (repeat cycle) associated with all polar orbiting satellites significantly limits their use for rip current identification. The repeat cycle for the SPOT optical satellite is 26 days. It may be lessened somewhat by directing the sensor, but it would still be impractical for use in a rip warning system.

Other types of satellite-based systems that may be able to observe rip currents directly are synthetic aperture radars (SARs) and interferometric synthetic aperture radars (INSARs). The spatial resolution of SAR systems is sufficient to observe small-scale current structures (e.g., ERRA-SARX has ~2 m resolution), but the long repeat cycles make them unsuitable for real-time warnings.

SHORE-BASED MICROWAVE SYSTEMS

Shore-based microwave systems presently under development for nearshore hydro-dynamic applications (Farquharson et al., 2005) offer significant promise for direct observation of rip currents. They typically operate in the marine navigation band (X-band) and may be based on commercial marine radars. Navigational microwave systems are incoherent; therefore, no phase information can be extracted and returned signal power is the only measured quantity.

Other specialized radar systems that operate in the microwave band are coherent (Figure 8.4) and capable of directly resolving flow velocities in the nearshore via Doppler processing (Figure 8.5). These instruments potentially have the most direct application for real-time rip current observation. The typical range of operation of both types of systems is 2 to 5 km with 10-m resolution (Farquharson et al., 2005; Trizna, 2010). The systems could be deployed on piers, buildings, and even lifeguard stations to observe local wave fields (Trizna, 2010; Lyzenga et. al., 2010). Radar systems offer two primary benefits when implemented as a part of a comprehensive beach safety program. They can continuously monitor a large surf area and increase

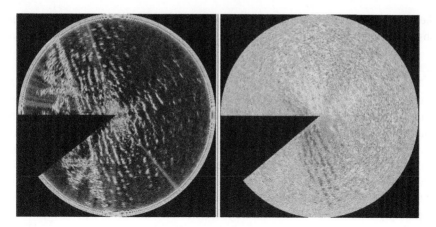

FIGURE 8.4   Coherent-radar derived images of surface reflectivity in the surf zone at Duck, North Carolina: (a) radar backscattered intensity image and (b) phase-difference image. (Courtesy of D. Trizna of Imaging Sensing Research.)

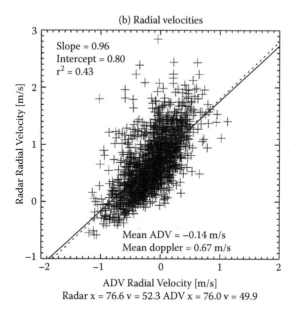

FIGURE 8.5   Comparison of radar velocity estimates with an ADV in the surf zone. (*Source:* Farquharson, G., S.J. Frasier, B. Raubenheimer et al. 2005. *J. Geophys. Res.*, 110: C12024, doi: 10.1029/2005JC003022. With permission.)

the amount of data available, and thus improve lifeguard training by validating visual observations of rip currents with direct measurements.

CONCLUSIONS

A wide range of remote sensing technologies may be applied for prediction and observation of rip currents. Local morphology, wave climate, and resources will dictate the suite of remote sensing, modeling, and/or in situ observations to be utilized. Particularly exciting are ground-based technologies such as HF radar for improved wave forecasts and marine-band radars for direct observation of rips. Their routine use will require additional research and coordination with beach safety personnel to determine the optimum technologies for specific coastal regions. These promising technologies will hopefully provide better and more timely information to the general public.

REFERENCES

Barrick, D.E. 1972. *Remote Sensing of Sea State by Radar; Remote Sensing of the Troposphere,* Derr, V.E., Ed., U.S. Government Printing Office, Washington, Chap. 12.

Bentamy, A., K.B. Katsaros, W.M. Drennan, and E.B. Forde. 2002. Daily Surface Wind Fields Produced by Merged Satellite Data. In *Gas Transfer at Water Surfaces*, Eds, M.A. Donelan, W.M. Drennan, E.S. Saltzman, and R. Wanninkhof, AGU, 343–349.

Brander, R.W. 1999. Field observations on the morphodynamic evolution of low wave energy rip current system. *Mar. Geol.*, 157: 199–217.

Brander, R.W. and A.D. Short. 2001. Flow kinematics of low-energy rip current systems. *J. Coast. Res.*, 17: 468–481.

Churnside, J.H. 2008. Polarization effects on oceanographic LIDAR. *Optics Expr.,* 16: 1196–1207.

Farquharson, G., S.J. Frasier, B. Raubenheimer et al. 2005. Microwave radar cross sections and Doppler velocities measured in the surf zone. *J. Geophys. Res.*, 110: C12024, doi: 10.1029/2005JC003022.

Gibeaut J.C. 2003. LIDAR: mapping a shoreline by laser light. *Geotimes*, 48: 22–27.

Gurgel, K.W., G. Antonischki, H.H. Essen et al. 1999. Wellen radar (WERA): a new ground-wave HF radar for ocean remote sensing. *Coast. Eng.*, 37: 219–234.

Haus, B.K. 2007. Surface current effects on the fetch limited growth of wave energy. *J. Geophys. Res.*, 112: C03003, doi: 10.1029/2006JC003924.

Haus, B.K., P. Work, G. Voulgaris et al. 2010. Wind speed dependence of single site wave height retrievals from phased-array HF radars. *J. Atmos. Oceanic Tech.*, 27: 1381–1394, doi: 10.1175/2010JTECHO730.1.

Haus, B.K., H.C. Graber, L.K. Shay et al. 2003. Along-shelf variability of a coastal buoyancy current during the relaxation of downswelling favorable winds. *J. Coast. Res.*, 19: 409–420.

Hisaki, Y. 1996. Nonlinear inversion of the integral equation to estimate ocean wave spectra from HF radar. *Radio Sci.*, 31: 25–39.

Holman, R.A. and J. Stanley. 2007. The history and technical capabilities of ARGUS. *Coast. Eng.*, 54: 477–491.

Howell, R. and J. Walsh. 1993. Measurement of ocean wave spectra using narrowbeam HF radar. *J. Oceanic Eng.*, 18: 296–305.

Irish, J.L., J.K. McClung, and W.J. Lillycrop. 2000. Airborne LIDAR bathymetry: the SHOALS system. *PIANC Bull.*, 103: 43–53.

Lipa, B. 1978. Inversion of second-order radar echoes from the sea. *J. Geophys. Res.*, 83: 959–962.

Lipa, B.J. and D.E. Barrick. 1986. Extraction of sea state from HF radar sea echo: mathematical theory and modeling. *Radio Sci.*, 21: 81–100.

Lippmann, T.C., and R.A. Holman. 1989. Quantification of sand-bar morphology: a video technique based on wave dissipation. *J. Geophys. Res. Oceans*, 94: 995–1011.

Lyzenga, D., O. Nwogu, and D. Trizna. 2009. Ocean wave field measurements using coherent and non-coherent radars at low grazing angles. IGARSS, Honolulu.

MacMahan, J., A.J.H.M. Reniers, E.B. Thornton et al. 2004a. Infragravity rip current pulsations. *J. Geophys. Res.*, 109: C01033, doi: 10.1029/2003JC002068.

MacMahan, J.H., A.J.H.M. Reniers, E.B. Thornton et al. 2004b. Surf zone eddies coupled with rip current morphology. *J. Geophys. Res.*, 109: C07004, doi: 10.1029/2003JC002083.

MacMahan, J.H., E.B. Thornton, and A.J.H.M. Reniers. 2006. Rip current review. *Coastal Eng.*, 53: 191–208.

Marmorino, G.O., L.K. Shay, B.K. Haus et al. 1999. An EOF analysis of HF Doppler radar current measurements of the Chesapeake Bay buoyant outflow. *Cont. Shelf Res.*, 19: 271–288.

Mehra, A. and I. Riven. 2010. A real-time ocean forecast system for the north Atlantic Ocean. *Terr. Atmos. Ocean. Sci.*, 21: 211–228 .

Nieto-Borge, J.C., G.R. Rodriguez, K. Hessner et al. 2004. Inversion of marine radar images for surface wave analysis. *J. Atmos. Ocean. Technol.*, 21: 1291–1300.

Plant W.J., W.C. Keller, and K. Hayes. 2005. Simultaneous measurement of ocean winds and waves with an airborne coherent real aperture radar. *J. Atmos. Ocean. Technol.*, 22: 832–846.

Reniers, A.J.H.M., J. MacMahan, E.B. Thornton et al. 2006. Modelling infragravity motions on a rip-channel beach, *Coastal Eng.*, 53: 209–222.

Reniers, A.J.H.M., J. MacMahan, E.B. Thornton et al. 2007. Modeling of very low frequency motions during RIPEX. *J. Geophys. Res.*, 112: C07013, doi: 10.1029/2005JC003122.

Robertson W.V., D. Whitman, K.Q. Zhang et al. 2004. Mapping shoreline position using airborne laser altimetry. *J. Coastal Res.*, 20: 884–892.

Shay, L.K., J. Martinez-Pedraja, T.M. Cook et al. 2007. High frequency radar mapping of surface currents using WERA. *J. Atmosph. Oceanic Technol.*, 24: 484–503.

Shay, L.K., H.E. Seim, D. Savidge et al. 2008. High frequency radar observing systems in SEACOOS, 2002–2007: lessons learned. *Marine Technol. Soc. J.*, 42: 55–67.

Splinter, K.D. and R.A. Holman. 2009. Bathymetry estimation from single-frame images of nearshore waves. *IEEE Trans. Geosci. Remote Sens.*, 47: 3151–3160.

Tolman, H.L., B. Balasubramaniyan, L.D. Burroughs et al. 2002. Development and implementation of wind generated ocean surface wave models at NCEP. *Weather Forecasting*, 17: 311–333.

Trizna, D. 2010. Coherent marine measurements of properties of ocean waves and currents. IGARSS, Honolulu.

Voulgaris, G., B.K. Haus, P. Work et al. 2008. Waves initiative within SEACOOS. *Marine Technol. Soc. J.*, 42: 68–80.

Wyatt, L.R. 1990: A relaxation method for integral inversion applied to HF radar measurement of the ocean wave directional spectrum. *Int. J. Remote Sens.*, 11: 1481—1494.

Wyatt, L.R., S.P. Thompson, and R.R. Burton. 1999. Evaluation of high frequency radar wave measurement. *Coastal Eng.*, 37: 259–282.

Wyatt, L.R., G. Liakhovetski, H. Graber et al. 2005. Factors affecting the accuracy of Showex HF radar wave measurements. *J. Atmos. Oceanic Technol.*, 22: 847–859.

Wyatt, L.R., J.J. Green, and A. Middleditch. 2009. Signal sampling impacts on HF radar wave measurement. *J. Atmos. Oceanic Technol.*, 26: 793–805.

9 Effectiveness of Panama City Beach Safety Program

John R. Fletemeyer

CONTENTS

INTRODUCTION

Rip currents are powerful, channeled currents originating in the surf zone that extend varying distances offshore. Rip velocities vary from only 0.3 m/sec to more than 2 m/sec and are influenced by a number of variables such as tidal stage, wave height, wind speed and direction, and bottom topography. Structures such as piers, groins, and jetties represent areas where rip currents predictably occur (Figure 9.1). Natural rocky outcrops on beach faces and pocket beaches also create environments for permanent rips (Figure 9.2). Rip currents are commonly called rip tides (Figure 9.3) and undertows (Figure 9.4)—both terms are misnomers—and they exhibit some or all of the following attributes:

- A well-defined head and neck that has a mushroom shape at its terminus
- Flotsam and foam moving seaward in the current
- Discoloration in water because rip is more turbid
- "Excited" or active water in the rip current channel

Examinations of photographs and field evaluations of rip currents on many surf beaches worldwide indicate that they seldom resemble the description cited above. Consequently, public educational materials about rips are over-simplified and can

FIGURE 9.1   Strong rip current associated with man-made groin, San Juan, Puerto Rico.

FIGURE 9.2   Rip currents often form in areas where natural rock outcrops occur near beaches in Southern California.

be misleading because rips are not accurately and reliably depicted. This over-simplification may lead some bathers to believe that they can stand at the water's edge and spot rip currents; this may be possible in some cases, but certainly not in every case. Fletemeyer and Leatherman (2010) concluded that "expecting a bather to be able to identify rip currents using current signage and information is problematic, overly optimistic, and even unrealistic."

FIGURE 9.3   Incorrectly worded rip tide warning sign at Lauderdale-by-the-Sea, Florida.

FIGURE 9.4   Warning sign confusing undertow with rip current at Palm Beach, Florida.

Rip currents are estimated to claim 100 to 150 victims annually on U.S. beaches (Lushine, 1991). An objective for future research is to develop a more reliable reporting system that would necessarily involve medical examiners and assimilation of accurate information from both guarded and unguarded beaches. Considering that over 90% of U.S. beaches lack lifeguards, this is a daunting task. Another objective is identifying rip current "hot spots" where rips are common and a high incidence of drowning exists.

PANAMA CITY BEACH STUDY AREA

Panama City Beach (PCB), which is 43 km long, is located in Bay County on the Florida panhandle (Figure 9.5). It has 148,000 permanent residents and 6 million tourists visit this beach annually according to Dan Rowe, president of the Bay County Tourist Development Council. A new airport expansion project was recently completed to promote tourism, and the major draw is the beach.

PCB is characterized by a well-developed system of sand bars. The shoreline has a cusp-like configuration as influenced by transverse bars. During calm water and low tide, these distinctive bar features can be easily identified, especially from the high-rise beachfront hotels and condominiums (Figure 9.6). Seasonal water temperatures differ considerably, ranging below 60°F during the winter months to the mid and high 80s in the summer months (Figure 9.7). During the early spring, late fall, and winter, water temperatures are generally not conducive for recreational bathing.

Beachfront development includes single-family houses, stilt homes, town houses, motels, and high-rise condominiums and hotels. Pier Park has a large parking lot, making it a popular location for tourists and locals. The city has three public fishing piers and thirteen storm water outfalls empty onto the beach (Figure 9.8). During heavy rains, water flowing rapidly from these outfalls is responsible for cutting channels through the beach. Storm water outfalls may correlate with the locations of persistent rip currents, but this assertion must be investigated in the field. A number of dune walkovers to the beach display signs describing the meanings of warning flags posted on the beach (Figure 9.9).

FIGURE 9.5   Panama City Beach, Florida.

FIGURE 9.6 *(See color insert.)* Beach cusps at low tide in the Panama City Beach study area.

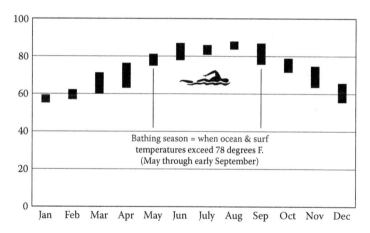

FIGURE 9.7 Panama City Beach monthly water temperatures in 2003.

DROWNINGS

Examination of the Bay County Medical Examiner's records over an 18-year period beginning in 1992 indicates a total of 153 salt water drownings. Regression analysis shows no statistical change in the drowning numbers over time (Figure 9.10). An epidemiological evaluation conducted by Fletemeyer (2005) utilizing PCB police reports of drowning events during a 2-year period beginning January 2002 revealed 69 drowning events involving 97 victims. Most of these events occurred during a

FIGURE 9.8   Storm water outfalls may contribute to the formation of rip currents on Panama City Beach.

3-month period (Figure 9.11). A total of 34 (49%) of the drowning events involved single victims, and approximately two-thirds of the victims were males. The average victim age was 30 (Figure 9.12); all were Caucasians, and tourists were involved in 83% of the drowning events.

SAFETY PROGRAM

Beach safety in Bay County is addressed in several ways and this section presents an overview of beach safety at PCB.

Beach safety officers and sheriff's deputies—Beach safety officers from the PCB Police Department are responsible for changing the warning flags and providing verbal warnings to the public about beach hazards, primarily about rip currents. They are directed not to perform rescues (Figure 9.13). At present, three beach safety officers are responsible for 27 km of beach. Considering the length of beach and scheduling around days off and vacations, this mission is problematic. In 2009, the city commission adopted an ordinance allowing sworn police officers to arrest bathers refusing to leave the water when double red flag warnings are flown. Members of the county sheriff's department receive water rescue training and are often called to effect rescues.

Lifeguards—During the past 20 years, seasonal lifeguards have occasionally been stationed on the beach. Currently, a single manned lifeguard tower is located at Pier Park. The management of the Holiday Inn Sunspree Hotel privately employs seasonal lifeguards (Figure 9.14). For liability reasons, they are called beach attendants—not lifeguards.

Beach warning flags—Flag poles are strategically located along the length of the beach. Four colors are used to represent different hazard conditions (Figure 9.9).

FIGURE 9.9   (*See color insert.*) Flag and rip current signs located at beach access points on Panama City Beach.

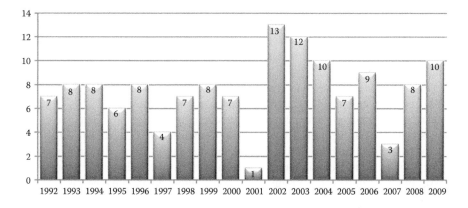

FIGURE 9.10   Yearly comparisons of the number of drownings in Bay County, Florida.

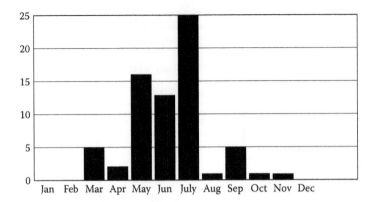

FIGURE 9.11   Monthly frequency of drownings in the Panama City Beach study area.

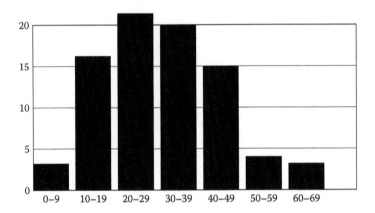

FIGURE 9.12   Age distribution of rip current drowning victims on Panama City Beach.

A double red flag represents an extreme hazard (water closed to the public). A single red flag denotes a high hazard of high surf and/or strong currents. A yellow flag indicates a medium hazard—moderate surf and currents. A green flag represents a low hazard and relatively calm conditions. A purple flag indicates the presence of dangerous marine life.

The effectiveness of the flag system has been challenged by Fletemeyer (2005). A flag represents a form of symbolic language and written information is required to clarify its meaning. Some people are functionally illiterate and hence may not understand the meanings of the flags. Bathers from different countries cannot always read or understand English. About 50% of the population is visually impaired and requires vision correction measures. These bathers commonly leave their corrective glasses or lenses in their hotel rooms or at home before going to the beach. In addition to these factors that limit the effectiveness of the warning flag system, flags are not always changed in a timely manner to accurately reflect ocean conditions and

FIGURE 9.13 Vehicle used to promote beach safety at Panama City Beach.

FIGURE 9.14 Beach safety personnel at the Holiday Inn Sunspree in Panama City Beach are technically not lifeguards.

hazard levels. The resulting confusion significantly impacts the effectiveness of the beach flag warnings.

Airplane banners—On particularly hazardous days when bathing loads are high, airplanes are chartered to fly banners over the beach to warn bathers about rip currents and other dangerous conditions. The effectiveness of this warning technique has not been evaluated.

Bather education program—With the aid of the local bed tax, the Bay County Tourist Development Council distributes a variety of beach safety materials including brochures, maps, and pamphlets and also maintains a website for tourists. Most hotels distribute beach safety information in their room directories. In addition to displaying information about the warning flags in each rentable lodging unit, signs explaining the flag colors are posted at beach access points, on billboards, and at various media and destination websites. In April 2009, an ordinance was passed requiring all lodging properties to prominently display interior signage detailing the flag warning system.

Text warnings—A text message warning program was implemented in 2008. Its objective is to issue alerts about changing surf conditions via texting.

Flag warning magnets—About 4,000 beach warning flag refrigerator magnets have been distributed to visitors through a partnership of the Panama City Beach Chamber of Commerce and the Gulf Coast Medical Center.

Other initiatives—With the goal of reducing the number of rip current drownings, various public education programs were implemented by the Bay County Tourist Development Council and local hotels and condominiums. The local American Red Cross chapter and Panama City Beach Community College promote beach safety in the Bay County area. The Southeast Region of the U.S. Lifesaving Association and the Florida Beach Patrol Chiefs Association have participated in two beach safety symposia funded by Florida Sea Grant.

BEACH SAFETY SURVEY

A 12-question survey instrument was designed to analyze beachgoer swimming abilities, the importance of having lifeguards, rip current knowledge, and familiarity with the flag warning system. During 10 days in August 1994, the beach safety survey was randomly administered to 130 bathers (62 males and 68 females), Randomization was accomplished by interviewing every tenth bather walking on the beach at the water's edge. Members of the Florida Beach Patrol Chiefs and the Southeast Region of the U.S. Lifesaving Association assisted. A follow-up study was conducted over an 11-day period beginning July 1, 2004. A total of 264 people (80 males and 184 females) were interviewed at the same location where the first survey was conducted. The Bay County Tourist Development Council participated. The 2004 survey included an additional question about the consumption of alcohol when visiting the beach. Table 9.1 presents the findings of the survey.

DISCUSSION

Tourists represented 90% of the beachgoers in both surveys. Swimming ability appeared nearly similar in the two surveys: 50 to 55% were fair swimmers and 9% were non-swimmers. The relatively high percentage of fair and non swimmers likely contributed to the large number of drownings at PCB.

TABLE 9.1
Survey Results from 1994 and 2004 Questionnaires (Percentages)

	1994	2004
What is your swimming ability? Good: can swim 1.6 km continuously.		
Fair: can swim at least 100 m (length of a football field) without stopping.		
Good	41	36
Fair	50	55
Can't swim	9	9
When going to the beach, is it important for a lifeguard to be present?		
Very important	59	40
Important	37	36
Not important	4	24
Do you sometimes consume alcohol on the beach?		
Yes	QNA	74
No	QNA	26
If you are staying at a local hotel or condo, were you given information		
about beach flags and rip currents?		
Yes	12	72
No	88	28
Are you familiar with the beach flag system and do you know the meaning		
of each of the flag colors?		
Yes	33	74
No	67	26
What are the meanings of the red, yellow, and purple flags?		
Correct response	25	58
Incorrect response	75	42
Of the five beach safety hazards, check the one you believe is responsible		
for causing more deaths and injuries to bathers.		
Lightning strike	12	5
Spinal injury resulting from dive into shallow water	0	6
Being caught in a rip current	32	50
Being attacked by a shark	24	13
Being trapped in an undertow or rip tide	32	26
If you observed a double red flag, how would you respond?		
Ignore it and continue to swim	8	4
Be more cautious and not venture far from shore	29	34
Stay out of the water	63	62

(continued on next page)

TABLE 9.1 (continued)
Survey Results from 1994 and 2004 Questionnaires (Percentages)

	1994	2004
If caught in a rip current, what would you do?		
Correct response	11	22
Partially correct response	60	53
Incorrect response	29	25
If you saw someone in a rip current, what would you do?		
Correct response	28	36
Incorrect response	72	64
If you were caught in a rip current, what should you do?		
Correct response	60	58
Incorrect response	40	42
Can you describe a rip current?		
Correct response	1	5
Incorrect response	99	95

QNA = question not asked (1994).

Bather awareness of rip currents and knowledge about identifying and escaping rip currents have increased overall. The study also indicates an increase in knowledge of the beach flag warning system. Despite this apparent increase in safety knowledge, the fact that the rip drowning rate has not been decreased over time suggests that the PCB initiatives have not been successful.

In 1994, 32% of the PCB bathers surveyed were able to identify rip currents as the most significant threats to their safety. Although the percentage of bathers able to identify the rip current threat increased 18% by 2004, half do not understand this hazard. This translates to about three million beachgoers annually.

A significant percentage of the PCB bather population do not know the recommended technique for escaping from rip currents. Only 22% of the bathers in 2004 were able to describe accurately how to escape the grip of a rip. In addition, a significant percentage of the bather population in 2004, although greater than in 1994, could not respond correctly when asked what was the first thing to do if they observed someone in a rip current. The correct answer was to call a lifeguard or 911 to seek other help.

Two questions in particular illustrate and measure the level of knowledge that bathers have about the beach safety program. Question 5 asks bathers if they are familiar with the beach flag system and 74% answered in the affirmative. In the follow-up question requiring bathers to correctly describe the meanings of the four flag colors, only 58% provided correct answers. This indicates a significant discrepancy between what people think they know and actually know about the flags and hence calls into question the effectiveness of the flag warning system.

A number of issues are responsible for limiting the effectiveness of the flag warning system. Importantly, the flags sometimes are not changed in a timely manner to reflect conditions correctly; this is especially true during early mornings and late afternoons. The failures of the flags to reflect the appropriate levels of hazards have also been observed by beach vendors (Fletemeyer, 2005).

Another reason for the relatively high drowning rate at PCB relates to alcohol consumption—74% of respondents admitting to drinking on the beach. The Centers for Disease Control stated that alcohol is a contributing factor in approximately 50% of drownings. The relatively high percentage response to this question came as a surprise because of an ordinance prohibiting alcohol consumption on the beach and imposing fines for consumption.

Perhaps the most important reason why the rip current drowning rate has not declined (Figure 9.10) relates to the decision not to implement a comprehensive lifeguard program. For decades the establishment of a lifeguard service has been debated by public officials and PCB stakeholders. A beach safety study was commissioned by the Mayor's Office and the Bay County Tourist Development Council. Fletemeyer (2005) recommended establishment of a seasonal lifeguard service consisting of thirteen lifeguard towers and the transition of police beach safety officers into full-time lifeguard supervisors. Despite strong public support, this recommendation has not been implemented. The efficacy of professional lifeguards has been clearly established (Branche, 2001); the lack of lifeguards is the principal reason for the high drowning rate at Panama City Beach, Florida.

Fletemeyer and Leatherman (2010) stated that signs, pamphlets, and related information about rips, while useful, are not sufficient for the public to identify these life-threatening currents. Typically, a rip current is depicted on a warning sign as having a mushroom-shaped head. After studies of hundreds of rip photographs and results of dye current studies, it is clear that these dangerous currents seldom resemble the drawings and figures commonly depicted in beach safety literature. Consequently the belief that bathers can be taught to identify rip currents and render judgments about whether or not it is safe to swim is problematic. In fact, the case could be made that misjudgments by beachgoers (e.g., they do not see the mushroom shape and hence think that there are no rip currents) may be responsible for some drownings.

CONCLUSIONS

This study evaluated the effectiveness of the Panama City Beach safety program, especially regarding rip currents. While improvement has been noted in beachgoer hazard awareness, the drowning rate has not declined between 1992 and 2009. An ambitious marketing plan designed by Bay County to increase tourism will likely also increase the number of drownings. These findings indicate the importance of establishing a seasonal lifeguard program. The importance of lifeguards in reducing the number of drownings has been clearly established by the Centers for Disease Control (Branche, 2001).

REFERENCES

Branche, C. 2001. Lifeguard Effectiveness: A Report of the Working Group. Centers for Disease Control and Prevention, Atlanta, pp. 1–21.

Fletemeyer, J. 2005. Panama City Beach and Bay County, Florida Beach Safety Report. Commissioned by the Bay County Tourist Development Council, pp. 1–43.

Fletemeyer, J. and G. Wolfe. 1994. Beach Safety Symposium, Sea Symposium II, Florida Sea Grant Foundation, pp. 1–22.

Fletemeyer, J. and S. Leatherman. 2010. Rip currents and beach safety education. *J. Coastal Res.*, 26: 1–3.

Lushine, J.B. 1991. A study of rip current drownings and related weather factors. *Natl. Weather Dig.*, 16: 13–19.

COLOR FIGURE 1.5   Many open-coast fixed beach rip currents appear as "dark gaps" of "calm" water between areas of breaking waves. To inexperienced beachgoers, they often look like the safest places to swim. (Courtesy of Rob Brander.)

(a)

(c)

COLOR FIGURE 1.11   Time sequence of release of purple dye into a topographic rip current during a public demonstration at Tamarama Beach, Sydney, Australia. (Courtesy of Rob Brander.)

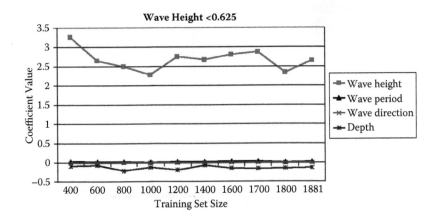

COLOR FIGURE 3.6 Plot of coefficients of linear regression formula for first branch of pruned tree model (wave height <0.625) versus sample size of training set.

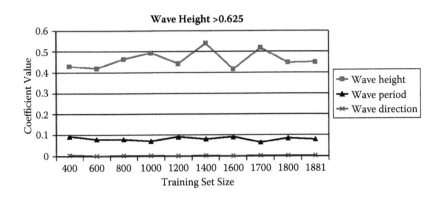

COLOR FIGURE 3.7 Plot of coefficients of linear regression formula for second branch of pruned tree model (wave height >0.625) versus sample size of training set.

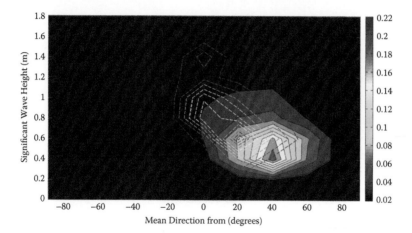

COLOR FIGURE 4.5 Contour plots showing bivariate distribution of significant wave height and mean direction of swells when only this partition is present for the entire data record (solid) and rip rescue record (dashed). A mean direction from 0 represents shore-normal incidence where negative degrees appear north of shore-normal and positive degrees south of shore-normal. Contour values are fractions of totals for each distribution.

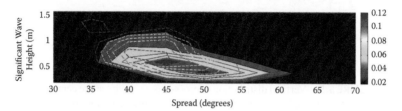

COLOR FIGURE 4.8 Bivariate distribution of directional spread and significant wave height of dominant swell for the entire data record (solid) and rip rescue record (dashed). Contour values are the fractions of total for each distribution.

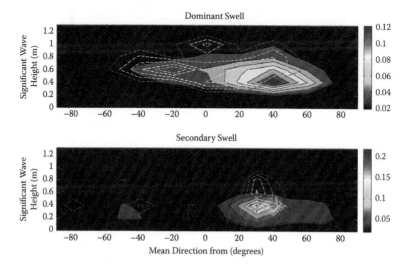

COLOR FIGURE 4.9 Top: contour plot showing bivariate distribution of significant wave height and mean direction for the dominant swell component. Bottom: secondary swell component. In both cases, the entire data record (solid) and rip rescue record (dashed) are shown; for the mean direction, 0 represents shore-normal. Data obtained only when two swells and no measurable wind sea were present.

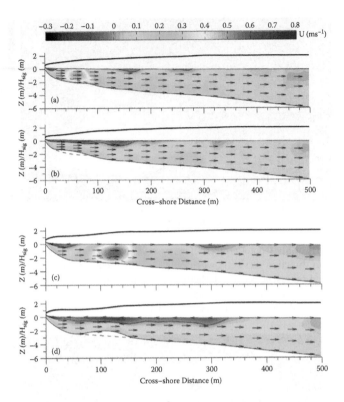

COLOR FIGURE 5.5 Vertical distribution of cross-shore velocity at the center of a rip channel (a and c) and bar (b and d) for bathymetric domains corresponding to Cases A (a and b) and B (c and d), respectively. Thick black lines show cross-shore profile of significant wave height. Locations of transects are shown in Figure 5.4.

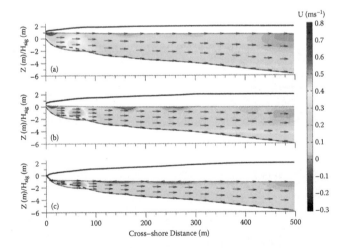

COLOR FIGURE 5.6 Case A: Cross-shore velocity (color contours) and significant wave height (black line) distribution along a transect through the rip channel location for three tidal stages: (a) high; (b) mean; and (c) low water levels. Thin black lines represent wave height distribution across the domain as estimated by SWAN. Note how the speed of the rip current increases with decreasing water level, while the offshore location of the maximum rip speed remains the same. The vectors indicate direction of flow; current strength is shown by the color-filled contours.

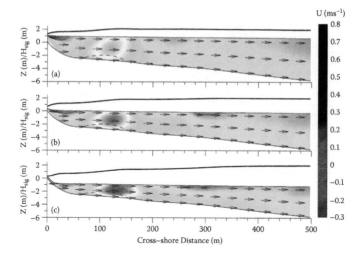

COLOR FIGURE 5.7 Case B: Cross-shore velocity (color contours) and significant wave height (black line) distribution along a transect through the rip channel location for three tidal stages: (a) high; (b) mean; and (c) low water levels. Thin black lines represent wave height distribution across the domain as estimated by SWAN. Note how the speed of the rip current increases with decreasing water level, while the offshore location of the maximum rip speed remains the same. The vectors indicate direction of flow; current strength is shown by the color-filled contours. The relative location of the bar is shown as a grey line.

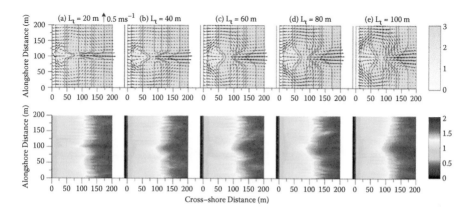

COLOR FIGURE 5.8 Circulation (depth-averaged current vector, top row) and significant wave height distribution (bottom row) for bathymetric domain of Case B with varying channel width (L_t) of (a) 20 m; (b) 40 m; (c) 60 m; (d) 80 m; and (e) 100 m.

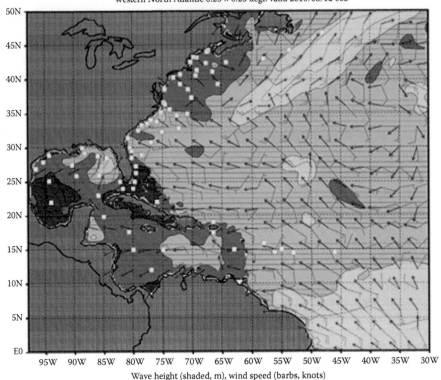

WNA 20100811 t06z 24h forecast
Western North Atlantic 0.25 × 0.25 degr. valid 2010/08/12 06z

Wave height (shaded, m), wind speed (barbs, knots)
and peak direction (vector, not scaled)
NOAA/NWS/NCEP Marine Modeling and Analysis Branch, 2010/08/11

0.5 1 1.5 2 3 4 5 6 7 8 9 10 11 12 13 14 15

(a)

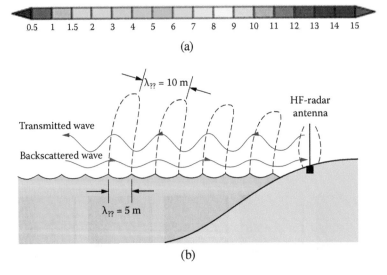

(b)

COLOR FIGURE 8.1   (a) Example of Western North Atlantic wave hindcast provided by NOAA/NWS/NCEP. (b) Principle of operation of HF radar: transmitted radar energy propagates over the air–water interface and is reflected back to the transmitter preferentially by waves of half the radar wavelength. (c) Sample wave spectrum derived from the University of Miami HF radar system using Seaview sensing software. (d) Modeled spectrum by SWAN utilizing local wind and current fields (from HF radar) as initial conditions. (Courtesy of Werner Gurgel, University of Hamburg.)

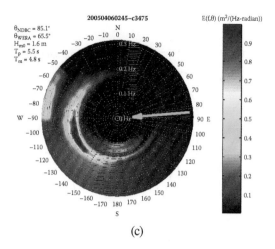

(c)

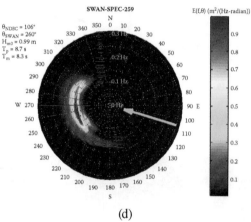

(d)

COLOR FIGURE 8.1 (continued).

COLOR FIGURE 9.6 Beach cusps at low tide in the Panama City Beach study area.

COLOR FIGURE 9.9 Flag and rip current signs located at beach access points on Panama City Beach.

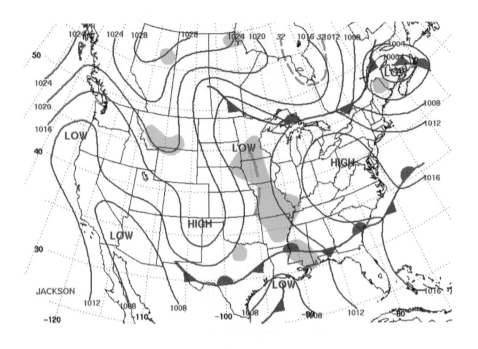

(a)

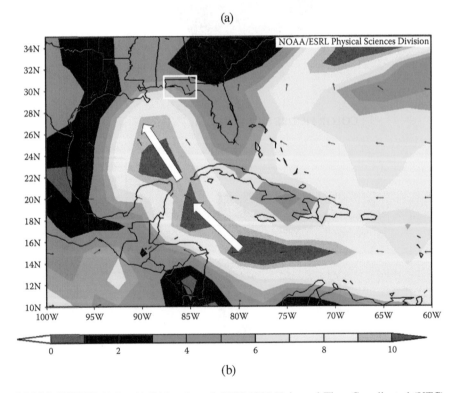

(b)

COLOR FIGURE 10.3 (a) Friday, June 6, 2003 1200 Universal Time Coordinated (UTC) surface weather chart. (b) 1200 UTC wind flow and speed (ms^{-1}).

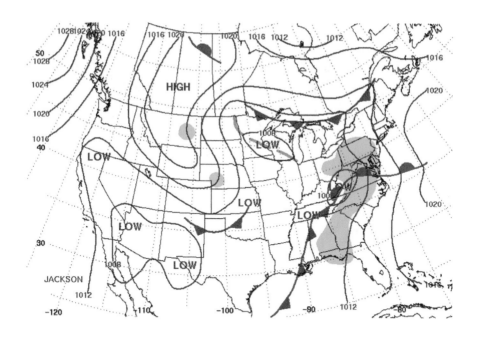

(a)

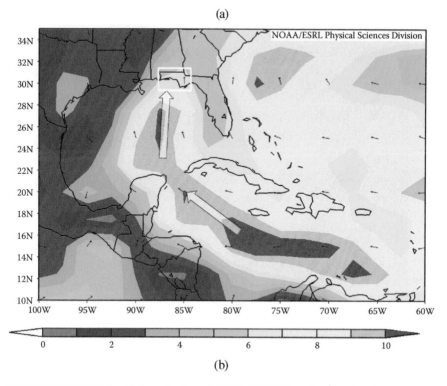

(b)

COLOR FIGURE 10.4　(a) Saturday, June 7, 2003 1200 UTC surface weather chart. (b) 1200 UTC wind flow and speed (ms^{-1}).

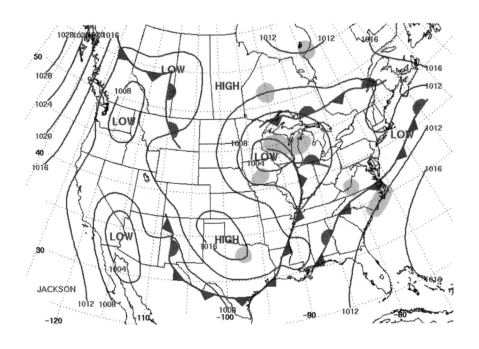

(a)

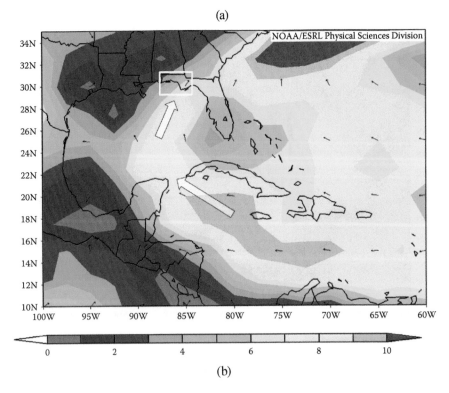

(b)

COLOR FIGURE 10.6 (a) Sunday, June 8, 2003 1100 UTC surface weather chart. (b) 1200 UTC wind flow and speed (ms^{-1}).

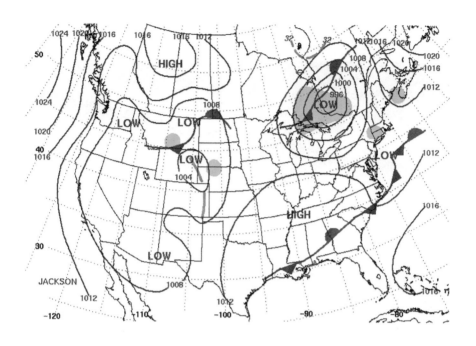

(a)

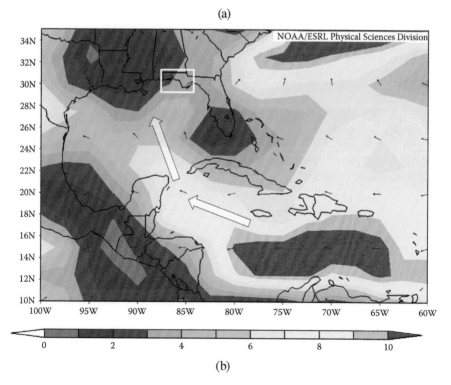

(b)

COLOR FIGURE 10.7 (a) Monday, June 9, 2003 1200 UTC surface weather chart. (b) 1200 UTC wind flow and speed (ms^{-1}).

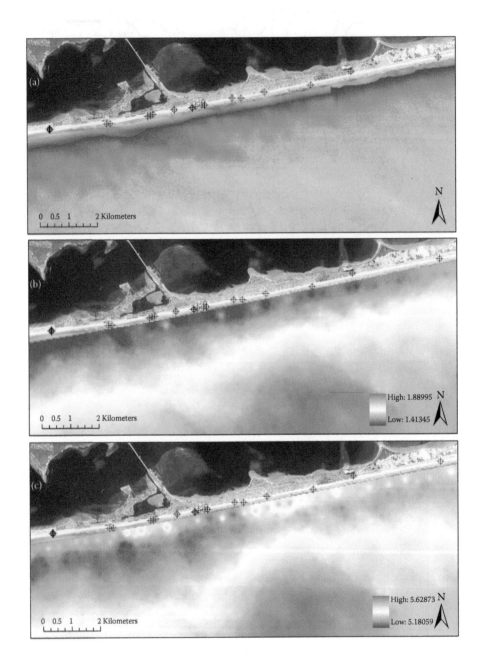

COLOR FIGURE 11.4 (a) Shelf bathymetry and drowning locations. (b) and (c) Representative wave model outputs. Shown are spatial variations of significant wave height (b) and mean wave period for wind and wave forcing (c) on March 15, 2008, when a drowning occurred at an unguarded section east of Casino Beach.

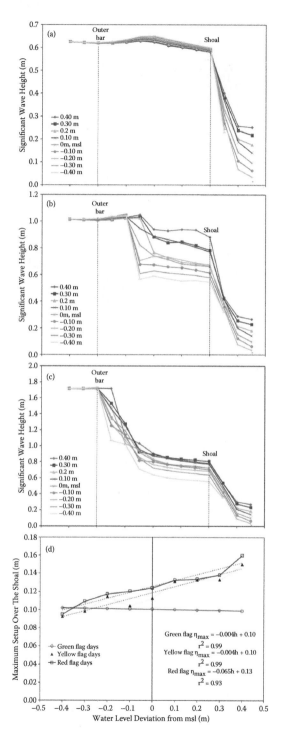

COLOR FIGURE 11.8 Modeled cross-shore transformation of incident wave fields for (a) green, (b) yellow, and (c) red flag conditions with respect to inner- and outermost bars; (d) is variation in maximum set-up across the shoal with respect to water level.

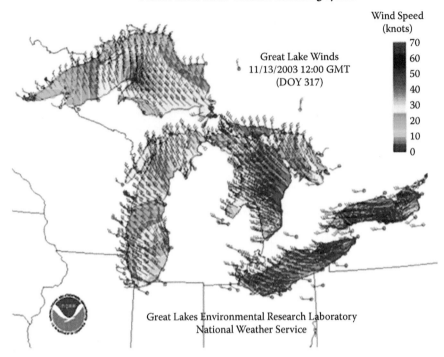

COLOR FIGURE 12.9 Strong, late season, outbreak of Canadian air over the Great Lakes Basin.

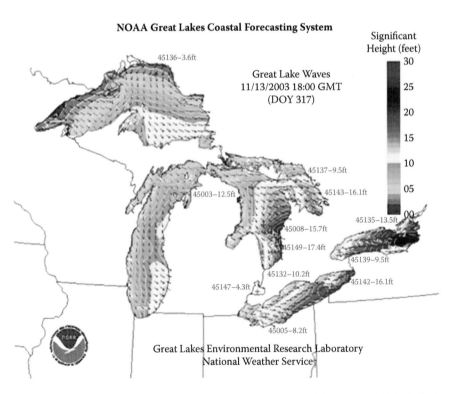

COLOR FIGURE 12.10 Resulting Great Lakes wave field (significant wave height in feet).

COLOR FIGURE 14.2 Crooklets Beach, north Cornwall provides an example of the complex nature of physical hazard dynamics. Low-tide rip systems, controlled by transverse bar, and rip beach morphology and upper beach morphodynamics, modified by geologic control and groundwater seepage, constrain and influence surf zone currents and hazards during mid and high tides. Inset shows location of southwest England study region (grey box) and Crooklets Beach location (solid circle). (Photo courtesy of Tim Scott.)

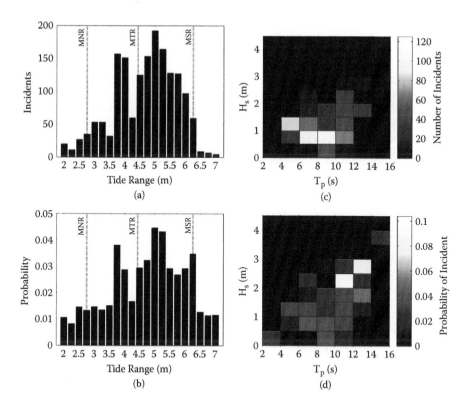

COLOR FIGURE 14.9 Histograms of tidal range associated with (a) incident frequency and (b) probability of incident (*IR*). Dashed lines indicate mean spring range (MSR); mean tidal range (MTR) and mean neap range (MNR) are marked. Data represent incidents recorded at all studied west coast LTT+R and LTBR beaches in 2007. Two-dimensional frequency matrices of joint wave distribution are associated with (c) number of incidents and (d) probability of incident. (*Source:* Data recorded during patrol hours, 2007 patrol season.)

COLOR FIGURE 16.4 Rip currents are sometimes controlled by underwater rocks and reefs. The "Shell Beach Express" is exposed at spring low tide on a rare low-wave day at La Jolla, California. The Shell Beach rip occurs in the channel between two rocks as delineated by the elongated body of light-colored sand that is swept out when the current is flowing. Many bathers have been pulled offshore in this powerful rip current during big surf conditions. (*Source:* Stephen P. Leatherman.)

COLOR FIGURE 16.7 Fluorescent green plumes from dissolving dye balls that move in a tidal current appear as white streaks in black and white. (*Source:* Stephen P. Leatherman.)

10 Meteorological Data Analysis of Rip Current Drowning

Charles H. Paxton

CONTENTS

INTRODUCTION

Shepard (1936) first used the *rip current* term to describe a circulation pattern of water from waves breaking on a beach with the return flow moving rapidly back out to sea through narrow channels in the surf zone. Longer period waves (>7 to 8 sec) break with more energy and can push water higher up the beach face, hence driving stronger rip currents (MacMahan, 2003). Two primary factors associated with rip current formation are longshore variations in bathymetry and varying height along the beach (Dalrymple, 1978). Lower tidal stages also play a role in rip strength (Brander and Short, 2001; Dronen et al., 2002).

A sandy beach typically exhibits an undulating sea bottom with shallower sand bars parallel to the beach separated by deeper troughs. The difference can be drastic on beaches with waist-deep water dropping off to well over head within only a meter or two. Poor swimmers who feel safe in the shallower water can be pulled offshore slightly by a rip and panic when their feet do not touch the sandy bottom. MacMahan (2003) analyzed 3 years of time-averaged video images at Duck, North Carolina, to study the stability and persistence of rip channels. He concluded that only large storms with strong longshore currents have significant impacts on bar morphology.

Gensini and Ashley (2010) analyzed rip current fatalities in the conterminous United States from 1994 through 2007 and found an average of 35 deaths each year that are more likely to occur during summer season weekends. Synoptic-scale meteorological conditions were investigated when rip current fatalities occurred and seven categories were created: (1) high pressure-created onshore winds, (2) low pressure-created onshore winds, (3) thunderstorms in the vicinity, (4) winds parallel to shore, (5) tropical systems, (6) onshore winds—other, and (7) no significant synoptic-scale features present. Gensini and Ashley found that 70% of all rip current fatalities were associated with surface high pressure systems that generated strong onshore winds and tropical cyclones. Previous research has shown that onshore winds are responsible for greater rip current frequency (Lushine, 1991). Furthermore, strong onshore winds create choppy, disturbed waves that are more likely to catch a swimmer by surprise because the tell-tale signs of rips are masked and rough conditions may also hide a person in distress from potential rescuers.

Mollere et al. (2001) examined regional wind, tides and swells that influenced suspected Florida Panhandle rip current deaths. He concluded that 94% of the 18 cases occurred when (1) wind was normal to the shore (2) tide was outgoing, and (3) average swell height was 0.7 m with 6- to 7-sec periods.

This chapter reviews the statistics of deaths and injuries from nearly 500 rip current reports and the attendant weather patterns along the contiguous United States oceanic coasts. The multiple drownings on Florida Panhandle beaches during the infamous "Black Sunday" is also investigated to analyze the meteorological conditions responsible for this horrendous event.

RIP CURRENT DEATH AND INJURY DATA

DATA COLLECTION

Rip current death and injury data from 1994 through 2009 were gathered from *Storm Data* reports (National Climate Data Center, 2010). Although the extent of injuries associated with rip currents can vary, an injury is typically listed in *Storm Data* when a near-drowning victim is taken to a hospital. *Storm Data* rip current records are input by National Weather Service (NWS) warning coordination meteorologists at the 27 coastal offices around the continental U.S. These meteorologists receive reports from emergency managers, law enforcement officials, and the media—primarily from newspaper clippings. The first rip current death and injury records were entered by certain NWS offices in 1994. Other offices have only provided records within the past several years. The rip currents fall within the *Storm Data* category of "Ocean and Lake Surf" events. Although rip deaths and injuries have been reported in the Great Lakes region, this area was not examined for this study. In Hawaii, the *Storm Data* list cites 25 combined ocean deaths since 1995 under the heavy surf or high surf categories without making distinctions for rip currents; thus the Hawaii reports were not used.

STATE STATISTICS

Table 10.1 indicates that Florida had far more rip current victims than any other state, with 234 deaths from 1994 to 2009. California was in second place with 43 deaths. North Carolina (36 deaths), Alabama (23 deaths), and New Jersey (22 deaths) round out the five states with the most rip current victims. Florida has 2,173 km of

TABLE 10.1

Continental States with Most Rip Current Deaths and Injuries (Storm Data 1994 to 2009)

State	Deaths	Injuries
Florida	234	199
California	43	97
North Carolina	36	14
Alabama	23	2
New Jersey	22	27

coastline, 1,067 km of beaches, and the warmest water of any state—reaching above 26°C during the summer and sometimes above 32°C in the shallow Gulf of Mexico waters. California has 1,352 km of coastline and much colder water. Average winter water temperatures in southern California according to National Oceanographic Data Center are around 14°C, dropping below 12°C in northern California. In summers, temperatures in southern California waters approach 21°C and 16°C in northern waters. North Carolina, New Jersey, and Alabama waters are cold during winters; therefore, most rip current incidents occur during the warmer months.

Table 10.2 shows the number of days with *Storm Data* reports of rip current deaths, injuries, or rescues. Winter is defined from November through April and summer as May through October. Along the east coast of the United States, only the heartiest swimmers go for dips when the water is cold. Most rip current deaths occur when water temperatures exceed 20°C. The exceptions are resort areas frequented by college spring breakers during March and April. Cold water decreases stamina and leads to muscle cramping and hypothermia. Intake of alcohol lowers inhibitions and increases the effects of cold water.

Alabama had four cases in April but none between November and March. The Florida Panhandle experienced 8 rip current days with deaths or injuries during March and 11 in April. The other exception was California, where water temperatures are usually chilly. In California, someone was killed or injured by rip currents during every month, with 21 case days from November through April.

Table 10.3 shows the Florida counties with the most rip victims. Escambia and Walton Counties are located in the Panhandle, and Miami-Dade and Broward Counties are in the Southeast (Figure 10.1). Not surprisingly, more rip deaths occurred over weekends and Sundays were the deadliest days (Figure 10.2).

SOCIOLOGICAL AND OTHER FACTORS

This study indicated that about 12% of rip current death victims were rescuers who drowned during lifesaving attempts. Without a flotation device, a panicking person caught in a rip current will try to use his rescuer as a flotation aid—with severe consequences. Around 90% of the rip current victims in this study were male, and many

TABLE 10.2

Number of Days with Rip Current Deaths, Injuries, or Rescues (Storm Data)

Location	Year of First Record	Winter: November–April	Summer: May–October	Total Rip Current Days
California	1995	21	39	60
Oregon	2007	2	1	3
Washington	2007	0	1	1
Texas	2007	0	6	6
Louisiana	2002	0	2	2
Mississippi	n/a	0	0	0
Alabama	1995	4 all April	19	23
Florida	1994	66	177	243
Florida Panhandle	1994	22	52	74
Florida Southwest	1997	3	9	12
Florida Southeast	1994	28	57	85
Florida East	1994	13	53	66
Georgia	2005	0	6	6
South Carolina	1997	1	11	12
North Carolina	1995	0	41	41
Virginia	1995	0	2	2
Maryland	2006	0	1	1
Delaware	1995	0	16	16
New Jersey	1995	0	30	30
New York	1998	1	11	12
Connecticut	n/a	0	0	0
Rhode Island	n/a	0	0	0
Massachusetts	n/a	0	0	0
Maine	n/a	0	0	0

n/a = Not available; no records.

TABLE 10.3

Five Florida Counties with Most Rip Current Deaths and Injuries (Storm Data 1994 to 2009)

County	Deaths	Injuries
Broward	32	41
Escambia	31	46
Miami-Dade	20	13
Walton	20	1
St. Johns	19	1

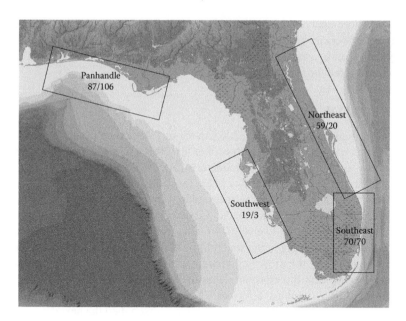

FIGURE 10.1 Florida coastal regions with rip current deaths and injuries.

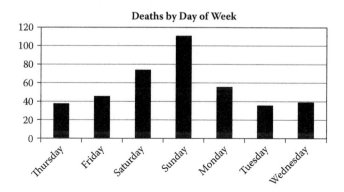

FIGURE 10.2 Rip current deaths by day of week, 1995 through 2009.

were tourists who were unfamiliar with ocean waters. Studying accidental drownings in Pinellas County Florida, Nichter and Everett (1989) found that bodies of salt water were the most common drowning sites; three times more males drowned than females, and 59% of young adult victims exhibited detectable blood alcohol levels. Only one of the 495 *Storm Data* reports used in this study mentioned alcohol usage. Copeland (1984) looked at Dade County drowning victims and found that almost 38% had detectable blood alcohol levels. Gulliver and Begg (2005) surveyed young New Zealand adults and found that males reported higher levels of water confidence, more exposure to risky behaviors, more exposure to unsafe locations, and more near-drowning incidents than females. They also determined that water-confident males were more likely to drink alcohol before water activities. Morgan et al. (2009) studied beachgoers in Australia and found that males visited surf beaches more frequently than females, spent more time in the water, ventured into deeper water, and more often entered the water after using alcohol.

BLACK SUNDAY

Eight People died on Sunday June 8, 2003, in the surf along the Florida Panhandle—a day now known as Black Sunday. Rowan et al. (2004), in a Florida Department of Health study, examined common characteristics of this event and other days that summer at Florida Panhandle beaches where four others died. The authors found that eight of twelve drowning victims in their study were male; ten of twelve were from out of state; and three of the twelve showed detectable levels of alcohol. The study also found that eight of the twelve (67%) who drowned were attempting to rescue people who were struggling in the water. Most of those struggling were later saved.

The median age of the drowning victims was 47 years old. Three of the twelve drownings occurred in the morning, and nine in the afternoon. One of those who drowned was Larry Lamotte, a former CNN correspondent. Black Sunday was a horrifying beginning of an ill-fated summer. One person died the next day, and four people had already died in Panhandle rip currents only a month earlier. Three others perished a month later, in July, bringing the total to sixteen deaths in Panhandle counties (Table 10.4). Four others died in Panhandle rip currents at the end of August and this brought the total deaths for the summer up to twenty. Accounts of the incidents in local time from Black Sunday and the next day from *Storm Data* (2003) provide more detail to this woeful story:

- **Okaloosa County (Destin), June 8, 2003, 1 PM**—A 57-year-old Missouri man was swimming in rough surf near James Lee Park when he started having trouble staying afloat. He was pulled out of the water but later died at a local hospital. Red flags were flying at the time of the drowning.
- **Okaloosa County (Destin), June 8, 2003, 3:15 pm**—A 31-year-old female from Indiana jumped into the Gulf to rescue her son who was having trouble swimming in the rough surf. Both were pulled out of the rough water and transported to a local hospital. She was pronounced dead, and her son was treated and released. The drowning occurred around James Lee Park where red flags warning of dangerous surf were flying. These drownings were part of a rash of drownings that occurred along the beaches of Walton and Okaloosa Counties on June 8. Six people drowned in Walton County that day. The rough water was the result of swells of 4 to 6 feet moving ashore. Ten people along the Okaloosa County beaches had to be transported to local hospitals where they were treated and released. Many others had to be pulled from the rough surf and were treated on the beach. Officials were flying and driving along the beaches to warn people of the dangerous surf. They would exit the water and then go back in after the officials left.
- **Walton County (Inlet Beach), June 8, 2003, beginning at 11:38 am**—At least twenty-eight people were rescued from rough surf off Walton County

TABLE 10.4
2003 Rip Current Deaths in
Florida Panhandle Counties

2003 Date	County	Deaths
May 9	Santa Rosa	1
May 9	Escambia	1
May 11	Escambia	1
May 31	Gulf	1
June 8	Walton	6
June 8	Okaloosa	2
June 9	Escambia	1
July 2	Bay	2
July 13	Bay	1
August 30	Escambia	2
August 31	Escambia	2
Total		**20**

beaches. Six persons drowned. As reported by Walton County Emergency Management, the victims were 60-, 53-, 40-, and 36-year-old males and 32- and 62-year old females.

• **Escambia County (Pensacola Beach), June 9, 2003, 1:58 pm**—A 66-year-old male from the local area jumped into the Gulf to rescue a youngster who was having trouble in the rough water. The child was brought ashore and survived. The 66-year-old drowned. Yellow flags were flying along the beach at the time of the drowning.

What were the conditions that killed those nine people on June 8 and 9, 2003? As with most misfortunes, a combination of events and not a single action led to this heart-breaking tragedy. The Florida Panhandle is a spring and summer destination known for its miles of white sand beaches that are accessible to tourists travelling from the Mid-West and Deep South. The water conditions on both June days were described as large swells moving onshore along the entire northwest Florida coastline. But that was only part of the story. The tragedies occurred over a weekend when the weather dictated peoples' actions.

Early that weekend, Friday, June 6, 2003 (Figure 10.3), a stationary front was draped over the land portion of the Gulf Coast. To the north were centers of high pressure over West Virginia with more vigorous high pressure sliding south just east of the Rocky Mountains. Low pressure just south of Louisiana created a southerly wind flow across the Gulf of Mexico. These strong and persistent winds, over 10 ms^{-1}, blew across the Caribbean Sea, northward through the Yucatan Channel between Mexico and Cuba, and across the Gulf of Mexico aimed just west of Florida toward Louisiana. Thunderstorms along the coastal Florida Panhandle during the afternoon put a damper on beach activities for early weekend visitors.

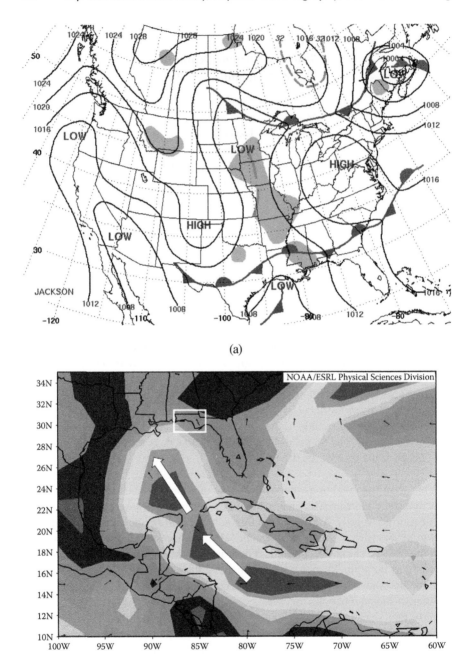

(a)

(b)

FIGURE 10.3 (*See color insert.*) (a) Friday, June 6, 2003 1200 Universal Time Coordinated (UTC) surface weather chart. (b) 1200 UTC wind flow and speed (ms^{-1}).

On Saturday, June 7, 2003 (Figure 10.4), the area of cooler high pressure sliding east of the Rockies pushed the west end of the stationary front eastward as a cold front, causing a massive rain area that dumped 50 to 100 mm of water along the Gulf Coast. The wind speed at Buoy 42036, 196 km west northwest of Tampa (Figure 10.5a), varied considerably from June 6 to 0000 UTC June 8 as rain squalls increased winds to over 8 ms^{-1}. Through these weather pattern changes, the southerly wind persisted over the Gulf of Mexico, creating building waves (Figure 10.5b).

On Black Sunday, June 8, 2003 (Figure 10.6), even though a complex arrangement of frontal systems was still in the area, the rain stopped. People were ready to play at the beach. The wind direction remained onshore but slowly trended from southeast to south southwest by 1200 UTC June 8. One element that changed little was the large choppy rough surf at the beaches. Buoy 42036 (Figure 10.5b) shows significant wave heights steadily building from Friday, peaking between 0600 and 1400 UTC on Sunday. From that point on, the seas began to subside but remained above 1 m during the day. The wave period through the event was in the 5- to 7-sec range and increased slightly from 6 to 8 sec as the wave heights peaked and began to subside.

On Monday, June 9, 2003 (Figure 10.7), although eight people drowned the previous day, sunny skies and warm temperatures did not inhibit people from enjoying another beach day. Unfortunately a 66-year-old man who tried to rescue a child in the surf drowned at the far west end of the Panhandle in Escambia County. The stationary front remained over the area, but the winds subsided. Most importantly, the seas had not completely diminished, remaining between 0.5 and 1 m in height (Figure 10.5b).

DISCUSSION AND CONCLUSIONS

The wind patterns before, during, and after Black Sunday in June 2003 were consistent with other days in which Panhandle rip current deaths occurred. What made this situation worse was a string of rainy weekend days that kept tourists cooped up in their hotel rooms. When the weather cleared, beachgoers had little sense of the dangers hidden in the aqua-green waters of the Gulf of Mexico.

Sadly, other multiple drowning events similar to Black Sunday occurred. Less than a month after Black Sunday, on the Fourth of July weekend, rainy and windy conditions on Saturday built significant wave heights up to 2 m with 7-sec wave periods. On Sunday, July 2, 2003, the skies temporarily cleared and the wind subsided, and two swimmers succumbed to rips in rough surf at Panama City Beach. Four people perished at Panhandle beaches on Sunday, August 3, 2008. Friday, August 1, and Saturday, August 2, were rainy with southerly winds up to 10 ms^{-1} that built seas up to 1.5 m with wave periods increasing from 5 to 7 sec. Again, a contributing factor was the rainy weekend weather that did not improve until Sunday.

Tropical cyclones generate large seas, including long-period swells that affect beaches near and far. Just 2 years before Black Sunday, Tropical Storm Allison created rough surf that killed five Florida Panhandle beachgoers on June 6 and 7, 2001. As slow-moving Allison made landfall near Houston, Texas on June 6, winds over the eastern Gulf of Mexico were southeasterly at 5- to 10 ms^{-1}, creating seas up to 1.7 m with 8-sec periods based on NOAA buoy data.

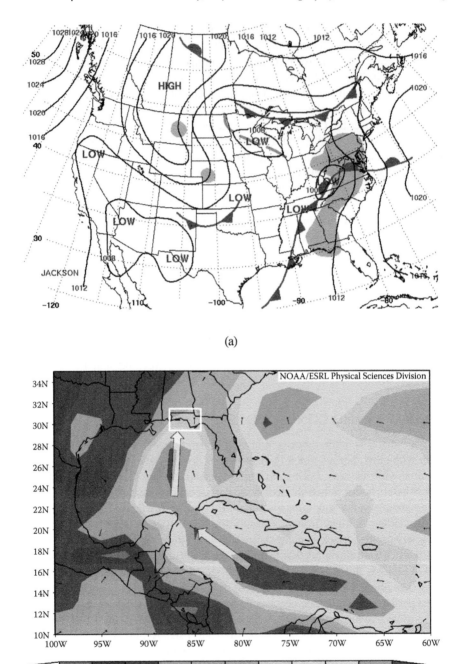

(a)

(b)

FIGURE 10.4 (*See color insert.*) (a) Saturday, June 7, 2003 1200 UTC surface weather chart. (b) 1200 UTC wind flow and speed (ms^{-1}).

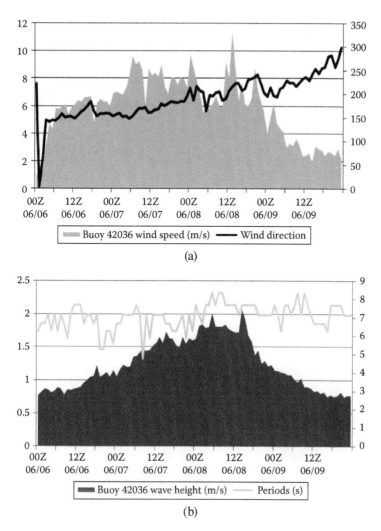

FIGURE 10.5 (a) Buoy 42036 wind direction and speed 196 km west northwest of Tampa, Florida. (b) Buoy 42036 wave height in (m) and period (sec) 196 km west northwest of Tampa. (Times are UTC.)

Many factors come into play for rip drowning, including beachgoer swimming abilities and surf experience. Other factors, such as the day of the week are important; more deaths occur when more people are at the beach. Sunny, warm weather also invites visits to the beach. Many victims are tourists who are unfamiliar with ocean forces, particularly rip currents. Most rip deaths occur in areas without lifeguards.

Education is one of the keys to safer beach trips. Many beach access points now have standardized signs that pictorially show how to survive a rip current. Such signage is helpful but not sufficient. Stronger wording, similar to that used at Ocean Beach in San Francisco, California may keep weaker swimmers out of the water. Perhaps more signs should state: Danger Rip Currents—People Have Drowned Here.

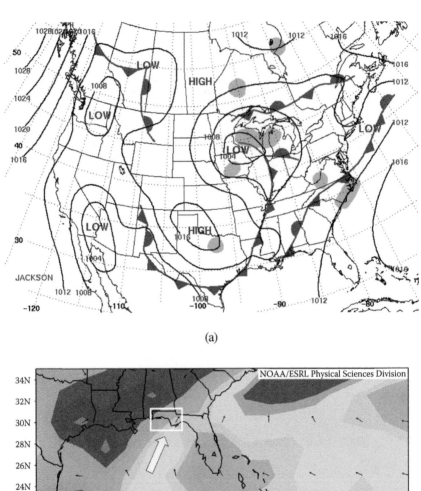

(a)

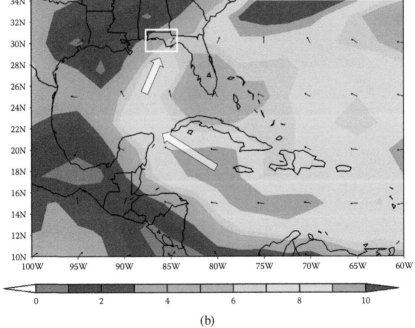

(b)

FIGURE 10.6   (*See color insert.*) (a) Sunday, June 8, 2003 1200 UTC surface weather chart. (b) 1200 UTC wind flow and speed (ms^{-1}).

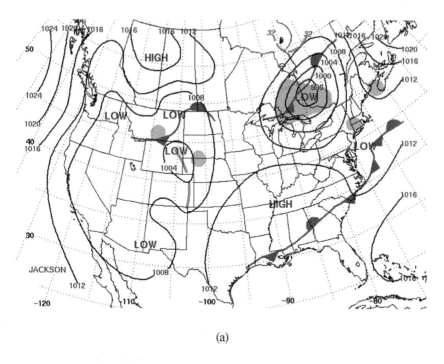

(a)

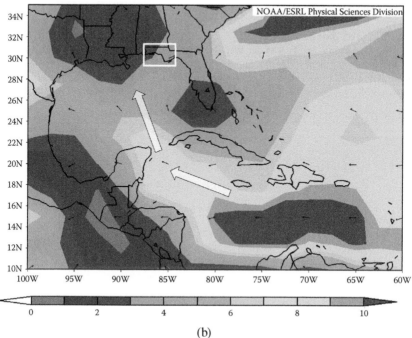

(b)

FIGURE 10.7   (*See color insert.*) (a) Monday, June 9, 2003 1200 UTC surface weather chart.
(b) 1200 UTC wind flow and speed (ms^{-1}).

REFERENCES

Brander, R.W. and A.D. Short. 2001. Flow kinematics of low-energy rip current systems, *J. Coastal Res.*, 17: 468–481.

Copeland, A. 1984. Deaths during recreational activities, *Forensic Sci. Intl.*, 25: 117–122.

Dalrymple, R.A. 1978. Rip currents and their causes, *Proc. 16th Conf. on Coastal Engineering*, Vol II. New York: ASCE, pp. 1414–1427.

Dronen, N., H. Karunarathna, J. Fredsoe et al. 2002. An experimental study of rip channel flow. *J. Coastal Eng.*, 45: 223–238.

Gensini, V.A. and W.S. Ashley. 2010. An examination of rip current fatalities in the United States. *Nat. Hazards.* 54: 159–175.

Gulliver, P. and D. Begg. 2005. Usual water-related behavior and "near-drowning" incidents in young adults. *Austral. New Zeal. J. Public Health*, 29: 238–243.

Lushine, J.B. 1991. A study of rip current drownings and related weather factors. *Natl. Weather Dig.*, 16.

MacMahan, J.H. 2003. Field observations of rip currents. PhD dissertation, University of Florida, Gainesville. Publication AAT 3096642. Available from: Dissertations & Theses: Full Text (accessed August 18, 2010).

Mollere, G.J., A.I. Watson, and R.C. Goree. 2001. A Rip Current Assessment of the Florida Panhandle Coastal Waters. NOAA Technical Memorandum NWS SR, 210.

Morgan, D., J. Ozanne-Smith, and T. Triggs. 2009. Self-reported water and drowning risk exposure at surf beaches. *Austral. New Zeal. J. Public Health*, 33: 180–188.

National Climate Data Center (NCDC). 2010. *Storm Data.* http://www4.ncdc.noaa.gov (accessed July 22, 2010).

National Data Buoy Center. 2010. NOAA/NDBC. http://www.ndbc.noaa.gov (accessed September 9, 2010).

Nichter, M.A. and P.B. Everett. 1989. Profile of drowning victims in a coastal community. *J. Fla. Med. Assn.* 76: 253–256.

Rowan A., D. Atrubin, and L. Van der Werf-Hourigan. 2004. Panhandle Beach Safety Study. Florida Department of Health, unpublished.

Shepard, F.P. 1936. Undertow: rip tide or rip current? *Science*, 84: 181–182.

11 Rip Current Hazards at Pensacola Beach, Florida

Chris Houser, Nicole Caldwell,
and Klaus Meyer-Arendt

CONTENTS

INTRODUCTION

Rip currents are approximately shore-normal channels of seaward-directed flows, driven by alongshore variations in wave height and associated variations in the mean water surface elevation. The resulting circulation cells play an important role in nearshore processes through offshore sediment transport and shoreline change (Shepard et al., 1941; McKenzie, 1958; Greenwood and Davidson-Arnott, 1979; Short, 1985; Smith and Largier, 1995; Aagaard et al., 1997; Thornton et al., 2007), but also represent significant hazards to beach users (Short, 1985; Lushine, 1991; Short and Hogan, 1994). Current velocities within a rip channel are on the order of 0.2 to 0.65 ms^{-1} in low-energy environments (Sonu, 1972 ; Bowman et al., 1988; Sherman et al., 1993; Smith and Largier, 1995; Aagaard et al., 1997; Brander, 1999; MacMahan et al., 2008), but can reach peak velocities in excess of 1 and up to 2 ms^{-1} in higher energy environments (Sonu, 1972; Brander and Short, 2000).

The potential for drowning or rescue depends on the presence of a strong rip current at a specific time and place and also on personal behavior including alcohol consumption, gender, age, swimming ability, panic, and exhaustion (Morgan et al., 2009; Gensini and Ashley, 2009). Wilks et al. (2007) suggest that a rip hazard also depends on whether beach users (particularly tourists) recognize and obey safety information and warning flags. At Pensacola Beach, an important tourist destination

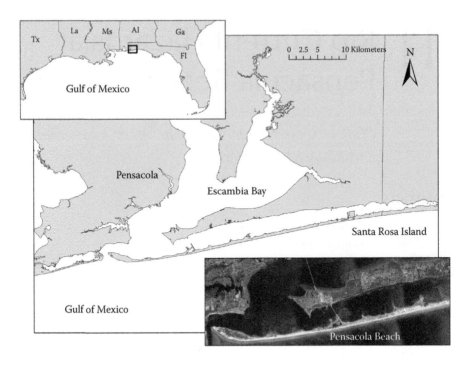

FIGURE 11.1 Location of Pensacola Beach in northwest Florida.

in northwest Florida (Figure 11.1), approximately 25% of drowning victims in the last decade were Good Samaritans trying to rescue others or parents going into the surf to rescue their children (Bob West, personal communication). Drownings and near-drownings usually cluster during summer weekends when strong onshore winds are associated with high pressure systems and large beach population are present (Gensini and Ashley, 2009).

Florida has the highest drowning rate in the United States (Lascody, 1998; Florida Department of Health, 2003), and the United States Lifesaving Association identified the Florida Panhandle as the nation's worst area for beach drownings. Drownings and near-drownings at Pensacola Beach tend to occur between March and October when the surf is strong (Lascody, 1998) and involve inexperienced swimmers who may not have heeded warning signs (Morgan et al., 2009; Fletemeyer and Leatherman, 2010).

Signage warns beach users of rip hazards at all access points along Pensacola Beach. Despite these clearly marked warning signs, beach users at Pensacola are unable to identify rip currents before entering the water and tend to swim in rip "hot spots" (Figure 11.2). A survey of 121 beach users during the summer of 2008 recreating at a persistent rip channel showed that 95% could not identify the active rip current. Additionally, only three of the thirteen respondents who had previously been caught in rip currents could identify the rip channel, even during yellow flag conditions when the rip channel was clearly defined by a breaking wave pattern. This may be due to the fact that the Florida Panhandle rips are more variable and appear closer to the shore (Figure 11.3) than the simple rip diagrams depicted on the warning signs.

(a)

(b)

FIGURE 11.2 (a) Representative photograph of accretion rip from perspective of a beach user showing eight people swimming in the rip channel and two children playing in the feeder channel that flows west to east. (b) Time stack (15-min) of same channel showing rip and complex topography of the bar and rip morphology.

Sonu (1972) examined a similar rip current system to the east of Pensacola Beach (Seagrove) that he described as alternating shoals and seaward-trending troughs. These accretion rips most commonly form during the recovery period following large storms as the shore face transitions from a dissipative to a reflective profile (Wright and Short, 1984; Short, 1985). The authors characterized the morphological state through a dimensionless fall velocity (Ω) given by:

$$\Omega = \frac{H_b}{\omega_s T} \tag{11.1}$$

where H_b is the breaking wave height, ω_s is the average fall velocity of the beach sediment, and T is the wave period. As wave energy decreases following a storm, the beach changes from dissipative ($\Omega > 6$) to reflective ($\Omega < 1$) by passing through each

FIGURE 11.3 Representative time exposure of rhythmic rip and shoal morphology at Pensacola Beach.

of four intermediate states: longshore bar–trough, rhythmic bar–beach, transverse bar–rip and low tide–terrace. Each intermediate state is characterized by the presence of rip currents, with the strength of the current in the feeder channel and neck varying in response to the landward migration and welding of the innermost bar. The current velocity increases as the cross-sectional area of the feeder channel decreases and the bar (or shoal) is more effective at blocking the return flow (Brander, 1999). As the feeder currents tend to be directly adjacent to the beach face (Figures 11.2 and 11.3), a beach user would not have to venture too far into the water before getting caught in rip channel and potentially needing assistance.

After development, rip channels induce an alongshore pressure gradient, leading to a strong current of 0.2 to 0.3 ms^{-1} across the shoal that contribute to feeder and rip current speeds of 0.4 to 0.6 ms^{-1} (Sonu, 1972; Mei and Liu, 1977). Over a rip and shoal system at Pensacola Beach, Houser et al. (in review) used a mass balance approach to estimate a depth-averaged current velocity of 0.2 ms^{-1} under green flag (low hazard, calm) conditions. While this velocity would not necessarily pose a significant hazard to an adult, it suggests that the velocities can be very strong during yellow and red flag days when wave breaking and set-up across the shoal would be more intense.

With increasing wave height, more water may enter a rip directly from the steep sides of the shoal rather than through the feeder channel (McKenzie, 1958; Sonu, 1972; Short, 2007), potentially leading to weaker currents near the shore. However, superimposed on the mean current are oscillatory currents at both gravity and infragravity frequencies that can more than double the seaward current (Sonu, 1972; Aagaard et al., 1997; Brander and Short, 2001; MacMahan et al., 2004). These currents can also create a pulsed flow, significantly contributing to drownings (Bob West, personal communication). Wave heights tend to be greater in rip channels than over shoals due to interactions with the seaward-flowing current and the deeper bathymetry (Sonu, 1972; Haller et al., 1998). This strong current can also cause seaward-migrating mega-ripples and possibly unstable footing that may be responsible for the "collapsing sand bar" phenomenon described by beach users as a sudden collapse of the bed and loss of footing (Short, 2007).

Rip currents at Pensacola Beach are morphologically controlled by transverse ridges on the inner shelf (Houser and Hamilton, 2009; Houser et al., 2008). The ridges are capable of refracting the incident wave field to create an alongshore varia-tion in wave height that causes an alongshore variation in the breaking wave height [Equation (11.1)]. Between ridges, where the breaking wave heights are smaller, the nearshore tends to be forced into a rhythmic transverse bar and rip state (TBR; Wright and Short, 1984) as shown in Figure 11.3, creating hot spots of rip current activity and drowning events. It is also possible that beach users are attracted to these areas because the surf is not as strong as along the ridge crests during yellow and red flag conditions—meaning that they may be unaware of the rips directly at the shore.

The TBR morphology develops soon after a reset storm detaches the innermost bar from the beach face. As the nearshore recovers from the storm, it progresses through a range of intermediate beach states, whereby the innermost bar becomes more crescentic and eventually attaches to the shoreline, and topographically forced rips develop rhythmically along the shoreline (Turner et al., 2007). Due to wave refraction over the transverse ridges, the rhythmic bar and rip state and the rip channels form first and persist longest between ridges, but will eventually be replaced by a nearshore terrace. The TBR beach state will develop nearer the (higher energy) ridge crests if low energy recovery conditions are maintained for a sufficiently long period.

The Santa Rosa Island Authority responsible for Pensacola Beach safety main-tains a flag system to alert beach users about heavy surf and rip hazards based on the National Weather Service forecasts. The highest flag color for that day (green, yellow, red, or double-red) is recorded by the authority, along with the number of prevents, assists, rescues, and contacts. Between March and October, lifeguards are stationed at Casino Beach, Fort Pickens Gate, and Park East; the remainder of the beach is patrolled by foot and vehicle. Lifeguards are permanently stationed at Casino Beach from March to October, and only between May and August at Fort Pickens Gate and Park East. Between 2004 and 2009, lifeguards performed 43,589 preventative acts, assisted 1,578 swimmers in distress, and rescued 759 swimmers who would have drowned without assistance.

Chi-square analysis suggests that prevents and assists are over-represented on red flag days and to a lesser extent on yellow flag days ($\chi2 = 681$, $p < 0.001$), while rescues tend to be more common on yellow flag days. Due to expanded beach patrols and the hiring of additional lifeguards in 2004 and 2005, only four drownings occurred—many fewer than the 21 drownings between 2000 and 2004. While most rescues concentrated within about 3.5 km of the central and popular Casino Beach, all the drownings occurred at unguarded sections of the beach or at times when lifeguards were not on duty; all occurred during red or double-red flag conditions. Rip cur-rents are identified by lifeguards and the media as the primary causes of drowning, although it is reasonable to expect that other factors (age, health, fatigue, heavy surf, etc.) also played a role.

While it has been demonstrated that drownings at Pensacola Beach tend to cluster between transverse ridges, little information is available about the current velocities during these events. The specific combinations of wind, wave, and tidal conditions causing drownings are not clear. A spectral wave and current model (MiKE21) was used to estimate the depth-averaged velocities and inshore wave heights of a representative rip and shoal system (Figure 11.2).

STUDY AREA

Development of Santa Rosa Island was initially limited by a lack of overland access and devastating hurricanes in the late nineteenth century, but fewer storms and the place-ment of permanent structures led to the construction of a bridge and the establishment of lifeguards at Casino Beach in 1931 (Meyer-Arendt, 1990). The recreational busi-ness began to develop following World War II as a small core of cottages, motels and other tourist-oriented businesses. Now commercial and residential development occu-pies about 13 km of shoreline and is bounded to the east and west by the Gulf Islands National Seashore.

Pensacola Beach receives 1.8 million visitors annually, in addition to the more than 270,000 day visitors and 670,000 visitors who stay at unpaid lodging sites. Toll data suggests that visits to the island are at a maximum during the summer holiday season (>400,000 vehicles per month) and at a minimum through the winter season (~200,000 vehicles per month). Most visitors indicate that the beach and its recreational oppor-tunities served as an "extensive influence" or "moderate influence" on their decisions to visit the area (Livingstone and Arthur, 2002). Visitors also indicate that cleanliness, climate, natural beauty, and beach safety are their most important concerns.

Based on a survey of beach users completed in 2008, most day visitors use the beach within a short distance of primary access points. Access is provided at the main Casino Beach (30.331006°N, 87.140842°W), Fort Pickens Gate (30.324764°N, 87.180515°W), and Park East (30.347154°N, 87.056521°W). Access is also possible via a walkover at "the cross" (30.329132°N, 87.153533°W) and at the ends of streets in the residential area east of Casino Beach, such as 23rd Avenue (30.339142°N, 87.097929°W). All the access points are between transverse ridges where rip cur-rents are common and drownings occurred (Figure 11.4). The high probability of rips and concentrations of beach users at these access points make them the most hazardous areas at Pensacola Beach.

METHODS

Since 2000, the Santa Rosa Island Authority has maintained a detailed rescue log listing the highest flag color for each day (green, yellow, red, or double-red) along with the number of drownings, prevents, assists, rescues, and contacts. To examine the wave and tidal forcing responsible for drownings, rescues and assists, a spectral wave and current model (MiKE21) was used to estimate the depth-averaged veloci-ties and inshore wave heights over a rip and shoal morphology (Wright and Short, 1984); this is representative of areas where drownings and most rescues and assists occurred at Pensacola Beach.

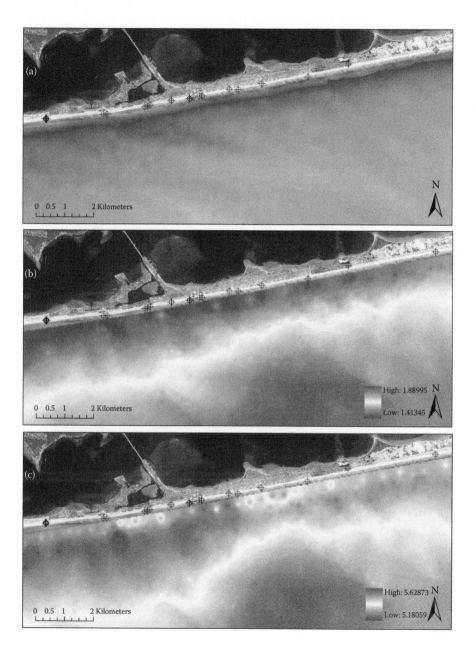

FIGURE 11.4 (*See color insert.*) (a) Shelf bathymetry and drowning locations. (b) and (c) Representative wave model outputs. Shown are spatial variations of significant wave height (b) and mean wave period for wind and wave forcing (c) on March 15, 2008, when a drowning occurred at an unguarded section east of Casino Beach.

The rip-and-shoal and nearshore morphologies within 200 m east and west of the largest transverse ridge was surveyed on June 16, 2008. While the specific morphology associated with each drowning, rescue, assist, and prevent differs, the results of this modeling analysis provide a first approximation of the conditions under which the events may have occurred. No observable changes in the morphology were noted within a week of the survey, and two rescues at this location involved use of U.S. Coast Guard helicopters to locate missing bathers. The survey was extended offshore using a fathometer to a depth of 15 m (~1.5 km offshore). The xyz survey data were interpolated to a flexible mesh with a maximum mesh area of 2 m² for a total of 61,184 nodes landward of the outermost bar.

The wave and current model was run using offshore wave boundary conditions and wind forcing function based on long-term meteorological and oceanographic records from two offshore wave buoys (42039 and 42040) located near the study region. A nested modeling approach was used to produce boundary conditions (both spectral and parametric) at the seaward extent of the modeling area where the finer grid resolution was used (Figure 11.5). Modeling results provide two-dimensional spectra for all boundary points in the study area (on a computational grid). The spectra were then used for finer scale transformations of the inshore area in order to analyze wave convergence and divergence across the shoal and rip. The model accounts for (1) wave propagation in time and space, shoaling, and refraction due to current and depth, (2) wave generation by wind, (3) wave–current and wave–wave interactions, (4) white capping, bottom friction and depth-induced breaking, and (5) wave-induced set-up and currents. The model utilized output from this wave transformation

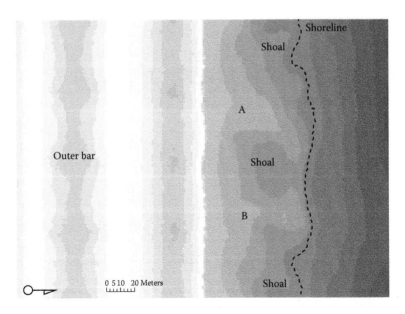

FIGURE 11.5   Interpolated bathymetry of the representative rip and shoal morphology used to model inshore wave heights and currents. This is the same rip and shoal complex shown in Figure 11.3.

module to generate currents resulting from the alongshore and cross-shore variations in wave set-up and set-down.

The inshore wave and current fields were generated for wave and tidal forcing at the time of each drowning (Table 11.1). Lifeguard records of times of drowning allowed the use of specific wave, wind, and tidal forcing in the model: the forcing for flag type, rescues, assists, and contacts was calculated as an average over daylight hours. The wave and current data were also generated for weighted average wave and tidal forcings for drownings, rescues, assists, prevents, and contacts (Table 11.2), for average green, yellow and red flag conditions, and for flag conditions weighted by the number of rescues.

RESULTS

DROWNINGS

Twenty-five drownings occurred at Pensacola Beach over a range of wave and tidal conditions between 2000 and 2009, and all were assumed by lifeguards and reporters to be associated with rip currents (Table 11.1). Some drownings were described as follows

- "Lifeguards went on duty over the weekend......but they arrived 5 days too late." (*Ocala Star Banner*, April 11, 2000)
- "Two men drowned about 2 hours after lifeguards went off duty.... She said the waves kept pulling her back out there.... They drowned in an area that has several drop-offs and sand bars where currents and waves can suddenly pull a swimmer into deep water...." (*Lakeland Ledger*, June 6, 2000)
- "Drowned while swimming...in an unguarded area under [yellow] surf warnings." (*Lakeland Ledger*, May 12, 2002)
- "[Tropical Storm] Hannah was presumed to have claimed the life of a body surfer...after getting caught in a rip current.... No lifeguards were on duty, but red flags and signs were posted, warning people to keep out of the water due to high surf and rip currents." (*USA Today*, September 15, 2002)
- "Rip currents apparently swept away a 19-year-old man wading in waist-deep water... [with] yellow caution flags but no lifeguards." (*Daily News*, March 29, 2003)

In other cases, reported rip drownings appear to have occurred during rough surf capable of toppling bathers:

- "Drowned in an unguarded area while swimming in rough conditions... in 6- to 8-ft waves kicked up by Tropical Storm Gordon." (*The Advocate*, September 22, 2000)
- "Being roiled by Tropical Storm Allison...several firefighters tried to reach Bill Thomas but they too were swept away by riptides." (*Sarasota Herald Tribune*, June 9, 2001)

TABLE 11.1
Incident Forcing for Submersions at Pensacola Beach, 2000 to 2008

Date	Location	Time	Tide Direction	Tidal Range (ft)	Water Depth (m)	Wind Direction	Wind Speed (ms⁻¹)	Wave Height (m)	Wave Period (sec)	Wave Direction (°)	Atmospheric Pressure (bar)
4/3/00	Ave. 21	—	—	—	0.15	148.0	10.4	1.6	6.7	1014.2	
6/4/00	FPG	1600	Out	2.5	0.05	238.6	2.9	0.4	6.0	1017.8	
6/4/00	FPG	1830	Out	2.5	0.12	229.4	3.5	0.4	6.4	1017.5	
9/17/00	Casino Beach	1930	In	0.8	0.25	41.7	9.3	3.7	9.0	1011.3	
9/20/00	Beach Resort	1930	In	1.5	0.05	122.7	6.7	1.0	5.0	1014.2	
10/27/00	Casino Beach	1330	In	1.0	0.13	—	—	—	—	—	
4/3/01	Five Flags	1330	In	1.5	0.10	160.7	7.5	1.3	5.2	153.4	1018.3
6/6/01	304 Ariola	1441	Out	2.0	0.20	132.1	7.0	1.6	7.5	186.6	1017.2
12/17/01	Clarion	1130	In	2.0	−0.10	159.1	8.7	1.6	6.6	165.1	1016.2
5/10/02	FPG	1300	Out	1.0	0.15	113.6	3.5	0.7	6.2	173.4	1019.8
5/13/02	BWJ	1125	Out	1.8	0.30	182.9	4.1	1.1	7.0	150.9	1018.0
7/5/02	1000 Ariola	—	—	—	0.05	239.6	7.7	0.7	6.1	239.6	1016.1
9/14/02	WPE	—	—	—	0.50	147.4	11.6	2.9	8.3	191.1	1012.9

11/10/02	Emerald Isle	0900	Out	2.0	0.13	206.7	7.8	1.8	7.0	159.6	1012.2
3/27/03	Seahorse	1440	In	1.5	0.25	95.9	5.3	1.5	6.7	140.7	1008.5
4/7/03	Holiday Inn	1050	In	1.3	0.26	135.1	3.5	1.0	5.7	174.0	1015.2
5/11/03	BWJ	1700	In	0.5	0.15	185.0	4.6	1.0	5.8	192.0	1016.9
6/9/03	Sans Souci	1400	Out	0.5	−0.03	260.1	3.1	0.8	7.1	187.9	1015.5
8/30/03	Emerald Isle	1230	In	0.3	0.38	100.9	6.6	1.6	8.7	164.0	1018.2
8/30/03	Portofino	1400	In	0.3	0.33	95.4	6.5	1.5	8.8	162.6	1018.6
8/31/03	Ave. 14–16	1030	Out	0.5	0.28	97.9	7.3	1.5	6.4	160.9	1017.2
3/15/05	BWDI	1623	In	2.2	0.15	—	—	1.0	4.4	30.7	1014.5
9/25/05	S of SYR	1623	Out	1.8	0.21	141.3	8.0	1.8	9.0	246.1	1017.5
9/14/08	PL 18 E of FPG	1620	Out	1.0	0.33	186.0	3.5	1.3	8.7		1016.1
9/14/08	PL 18 E of FPG	1800	Out	1.0	0.25	188.6	2.9	1.2	8.2		1015.7
Average				1.3	0.19	157	6.2	1.4	6.9	169	1016.0

Note: FPG = Fort Pickens Gate. BWJ = Beachside Windjammer. WPE = West of Park East. BWDI = Best Western Days Inn. SYR = Sabine Yacht & Racquet Club.

TABLE 11.2

Average Incident Wave, Wind, and Tidal Forcing for Drownings, Rescues, Assists, Prevents, and Contacts

	Speed (ms⁻¹)	Wind Direction (°)	Height (m)	Period (sec)	Water Level (m)
Drownings	6.21 ± 0.94	157 ± 8	1.41 ± 0.19	6.90 ± 0.71	0.19 ± 0.04
Rescues	6.70 ± 0.16	154 ± 7	1.16 ± 0.06	5.46 ± 0.21	0.04 ± 0.02
Assists	6.12 ± 0.09	150 ± 5	0.97 ± 0.04	5.18 ± 0.16	0.06 ± 0.01
Prevents	6.72 ± 0.02	152 ± 1	1.30 ± 0.01	5.97 ± 0.04	0.07 ± 0.002
Contacts	5.88 ± 0.01	171 ± 1	0.81 ± 0.003	4.54 ± 0.01	0.08 ± 0.0005

Note: Weighted by the number of each event.

- "Got caught in the surf...." (*The Advocate*, December 19, 2001)
- "Disappeared after she was hit by a wave in waist-deep water." (*South Florida Sun*, May 22, 2002)
- "Red flag conditions, meaning swimming is extremely dangerous.... Lifeguards said they rescued [an additional] 15 swimmers caught in rip currents off the guarded beach." (*USA Today*, August 31, 2003)

In some cases, Good Samaritans drowned while trying to rescue others:

- "Would be rescuers died trying to save children from drowning.... Both children were rescued by other beachgoers. ... apparently drowned Sunday while going to the aid of his granddaughter who was struggling in rough surf....died when he tried to rescue a 10-year old boy.... Neither [area] had lifeguards." (*Fredericksburg Free Lance Star*, April 8, 2001)
- "Drowned while trying to help other swimmers who were struggling with rough surf... under a yellow flag at the time." (*Gainesville Sun*, November 12, 2002)
- "Georgia man drowned Monday while trying to rescue his 12-year-old son struggling in rough surf... plunges into 3-ft waves to try to reach children.... The red flag at the end of the Pensacola Beach Gulf Pier warned of dangerous surf." (*Pensacola News Journal*, April 9, 2003)

The offshore significant wave height during these drownings averaged 1.4 m and ranged from 0.4 to 3.7 m (Table 11.3). The peak spectral wave period averaged 7 sec, with range from 4 to 9 sec. Water levels at the time of drowning averaged 0.2 m [above mean sea level (msl)], but ranged from –0.1 to 0.5 m. In all cases, the observed tide was greater than would have been expected from tide alone, with an average deviation of 0.1 m in response to storm winds from the south southeast. Bivariate plots of offshore significant wave height versus peak spectral period and significant wave height versus water depth are presented in Figure 11.6. There is no statistically significant difference in the wind and wave conditions between the periods when drowning did and did not occur.

TABLE 11.3
Predicted Significant Wave Height, Mean Current, Peak Oscillatory Current, and Combined Mean and Oscillatory Current for Shoal and Rip Channel over Representative Morphology for Each Submersion

Date	Location	Shoal				Rip Channel			
		H_s	C_{mean}	C_{osc}	C_{total}	H_s	C_{mean}	C_{osc}	C_{total}
4/3/00	Ave. 21	0.55	0.29	0.79	1.08	0.62	0.24	0.85	1.09
6/4/00	FPG	0.41	0.25	0.62	0.87	0.47	0.19	0.69	0.88
6/4/00	FPG	0.43	0.24	0.63	0.87	0.49	0.20	0.70	0.90
9/17/00	Casino Beach	0.76	0.37	1.09	1.46	0.86	0.33	1.11	1.44
9/20/00	Beach Resort	0.47	0.27	0.70	0.97	0.55	0.20	0.77	0.98
10/27/00	Casino Beach	No data available							
4/3/01	Five Flags	0.51	0.28	0.74	1.02	0.59	0.22	0.82	1.04
6/6/01	304 Ariola	0.56	0.28	0.80	1.09	0.63	0.25	0.86	1.10
12/17/01	Clarion	0.47	0.32	0.77	1.09	0.55	0.23	0.79	1.02
5/10/02	FPG	0.46	0.25	0.67	0.92	0.53	0.21	0.75	0.95
5/13/02	BWJ	0.54	0.25	0.74	0.99	0.61	0.23	0.83	1.06
7/5/02	1000 Ariola	0.43	0.26	0.66	0.92	0.50	0.20	0.72	0.92
9/14/02	WPE	0.77	0.30	1.00	1.31	0.86	0.31	1.08	1.39
11/10/02	Emerald Isle	0.56	0.30	0.82	1.12	0.64	0.25	0.87	1.12
3/27/03	Seahorse	0.57	0.27	0.79	1.06	0.64	0.24	0.87	1.11
4/7/03	Holiday Inn	0.53	0.25	0.72	0.97	0.60	0.22	0.82	1.03
5/11/03	BWJ	0.49	0.26	0.71	0.97	0.57	0.21	0.79	1.00
6/9/03	Sans Souci	0.41	0.28	0.67	0.95	0.47	0.21	0.70	0.91
8/30/03	Emerald Isle	0.61	0.27	0.82	1.09	0.67	0.26	0.89	1.15
8/30/03	Portofino	0.58	0.27	0.81	1.07	0.64	0.26	0.86	1.12
8/31/03	Ave. 14-16	0.58	0.27	0.79	1.06	0.66	0.24	0.88	1.12
3/15/05	BWDI	0.50	0.26	0.70	0.96	0.59	0.20	0.81	1.01
9/25/05	S of SYR	0.57	0.30	0.84	1.13	0.64	0.27	0.86	1.13
9/14/08	PL 18 E of FPG	0.56	0.26	0.78	1.04	0.62	0.25	0.84	1.09
9/14/08	PL 18 E of FPG	0.53	0.26	0.76	1.02	0.59	0.24	0.81	1.05
	Average	0.53	0.27	0.77	1.04	0.61	0.24	0.83	1.07

Note: Predicted significant wave height = H_s (m). Mean current = C_{mean} (ms^{-1}). Peak oscillatory current = C_{osc} (ms^{-1}). Combined mean and oscillatory current = C_{total} (ms^{-1}). FPG = Fort Pickens Gate. BWJ = Beachside Windjammer. WPE = West of Park East. BWDI = Best Western Days Inn. SYR = Sabine Yacht & Racquet Club.

Drownings are significantly over-represented for water levels greater than mean sea level and for tidal ranges larger than the mean. With the relatively small ranges and drownings equally distributed between incoming and outgoing tides, it is reasonable to expect that they are not associated with an additional current as the tide drains, but arise from a change in transformation of storm waves over the nearshore morphology. Model results suggest that for every 0.1 m increase in water level and

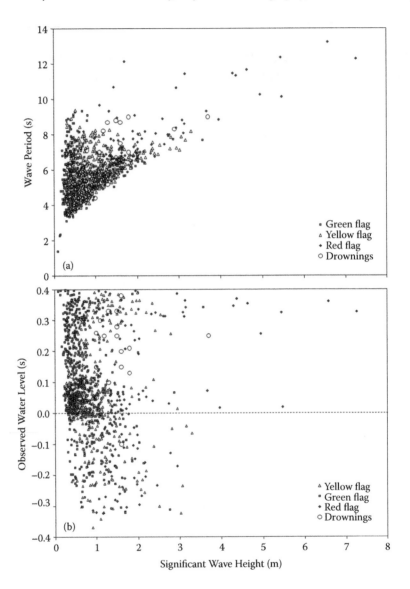

FIGURE 11.6 Scatter plots of (a) offshore significant wave height and wave period and (b) observed water levels for all flag types and for drowning events.

constant wave height and period, a 0.03 m increase in wave height appears along the seaward edge of the shoal. The 0.1 m increase in water level has a negligible effect on the inshore wave heights for green flag (small wave and period) conditions. An almost 0.05 m increase in inshore wave height creates red flag conditions (large wave and period). While seemingly small, a 0.05-m increase in wave height is enough to lift bathers and cause them to lose footing in a channel with depths of 1 to 1.5 m and both mean and oscillatory currents moving them seaward.

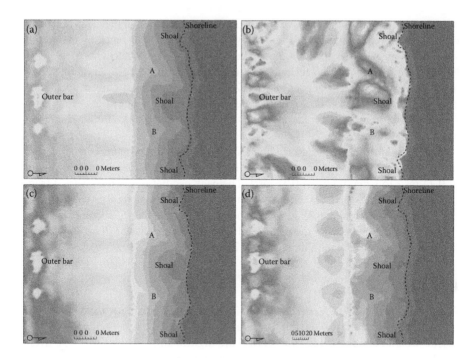

FIGURE 11.7 Spatial variations in (a) significant wave height, (b) mean current, (c) oscillatory current, and (d) maximum combined current speed for average of all drowning events at Pensacola Beach.

Model results showing spatial variations in mean current speed, peak oscillatory velocity, and significant wave height are presented in Figure 11.7 for average wind, wave, and tidal conditions during drowning events (Table 11.1). Significant wave height closely follows nearshore bathymetry (Figure 11.5), with larger waves closer to shore in the rip channels (A and B), consistent with the observations of Sonu (1972). Wave breaking through spilling across the shoal creates a wide zone of relatively small waves (<0.2 m) compared to the >0.3-m unbroken waves within the rip channel. Despite the greater depths of the rip channel, the larger waves are responsible for peak oscillatory velocities (>0.4 ms^{-1}) within the feeder current approximately 5 m of the shoreline (Figure 11.7c). Within 30 m of the shoreline, the peak oscillatory velocities exceed 0.6ms^{-1} at depths of 1 m. Comparatively, the mean current within the feeder current is only 0.1 ms^{-1} and increases to a maximum of 0.24 ms^{-1} approximately 50 m seaward of the shore. The landward-directed currents across the shoal reach a maximum of 0.28 ms^{-1} about 50 m from the shoreline, but remain 0.2 ms^{-1} along the seaward edge of the feeder channel. The maximum combined mean and oscillatory current is >0.4 ms^{-1} within 7 m of the shoreline and exceeds >0.7 ms^{-1} in the rip channel about 50 m from the shoreline.

The mean current in the rip channel ranged from 0.19 to 0.33 ms^{-1} over a wide range of incident wave and tidal conditions. In comparison, the peak oscillatory current within the rip channel varied from 0.69 to 1.11 ms^{-1} in response to significant wave heights of 0.47 and 0.86 m, respectively. The combination of the mean and peak oscillatory currents caused a pulsing flow within the rip channel, reaching maximum velocities between 0.88 and 1.44 ms^{-1}. The oscillatory current accounts for 78% of the maximum current in the rip channel—consistent with Sonu's observation (1972) of an oscillatory current accounting for 70% of the maximum current in the rip channel. This estimation does not include pulses in current speed at infragravity frequencies (Sonu, 1972; Aagaard et al., 1997; Brander and Short, 2000; MacMahan et al., 2003; Reniers et al., 2004).

The oscillatory current also accounts for 70% of the maximum current across the shoal for all of the drowning events. The mean current varied from a minimum of 0.24 ms^{-1} to a maximum of 0.37 ms^{-1}. The peak oscillatory currents varied from a maximum of 1.09 ms^{-1} to 0.62 ms^{-1} with maxima located along the seaward edge of the shoal. Within about 25 m of the shoreline, the oscillatory current was only 0.4 ms^{-1}, but combined with the mean landward current reached >0.6 ms^{-1} with broken wave heights of 0.2 m in depths of ~0.25 m.

RESCUES

Green flag rescues are generally associated with an offshore wave height of 0.70 m, wave period of 6 sec, and a tidal elevation of 0.04 m (Table 11.4). The rescue-weighted wave height and period are significantly greater than the average height (0.63 m) and period (5 sec) for all green flag days (with and without rescues), while the water level is significantly lower than the green flag average of 0.1 m. The mean current in the rip channel reaches a maximum of 0.18 ms^{-1} at a distance of 40 m from the shoreline. The inshore wave height is 0.46 m within the rip channel and 0.37 m along the seaward edge of the shoal, with peak oscillatory currents of 0.71 and 0.56 ms^{-1}, respectively. The combined current reaches a maximum of 0.89 ms^{-1} in the rip channel, which is significantly different at the 95% confidence level (p = 0.03) from the predicted current for all green flag days combined (0.85 ms^{-1}). With no statistically significant difference in the peak

TABLE 11.4

Average and Incident Wave Forcing[a] for Green, Yellow, and Red Flag Days, Mean Current, Oscillatory Current, Total Current, and Maximum Inshore Wave Height

Color	H_s (m)	T (sec)	H (m)	C_{mean} (ms^{-1})	C_{osc} (ms^{-1})	C_{total} (ms^{-1})	H_s (m)
Green	0.70	5.74	0.01	0.18	0.71	0.89	0.46
Yellow	0.93	6.58	0.04	0.23	0.68	0.91	0.49
Red	1.69	7.62	0.09	0.24	0.81	1.05	0.61
Green	0.63	5.31	0.10	0.14	0.72	0.85	0.51
Yellow	1.02	5.93	0.07	0.14	0.70	0.84	0.50
Red	1.73	6.97	0.05	0.15	0.69	0.84	0.48

[a] Weighted by number of rescues for flags with and without rescues.

oscillatory current and smaller wave heights on rescue days, it appears that green flag days with rescues are dependent on the strength of the mean current in the rip channel.

Yellow flag rescues are generally associated with an offshore significant wave height of 0.93 m, wave period of 6.6 sec, and water depth of 0.01 m. The wave period is significantly greater than all yellow flag days combined (5.9 sec), while the wave height and water depth are significantly lower than the wave height (1.02 m) and depth (0.07 m) for all yellow flag days combined. Despite smaller wave heights offshore, the combination of longer period waves in shallower water leads to greater dissipation across the shoal and a significantly stronger mean current of 0.23 ms^{-1} in the rip channel. The combined oscillatory and mean current reached 0.91 ms^{-1} within the rip current, although the oscillatory current alone was not significantly different from all yellow flag days combined (0.84 ms^{-1}). Similar to green flag conditions, yellow flag days with rescues appear to depend on the strength of the mean current in the rip channel.

Red flag days with rescues tend to be associated with offshore wave heights of 1.69 m, wave periods of 7.6 sec, and water depths of 0.09 m. The wave period and tidal elevation are significantly different from all red flag days (5.9 sec and 0.05 m), and the offshore wave height is significantly below 1.73 m for all red flag days combined. As a result of smaller wave height and larger tidal elevation, wave height in the rip channel was significantly larger at 0.61 m, leading to a greater oscillatory current. While the mean current in the rip channel was significantly larger than for all red flag days (0.24 versus 0.15 ms^{-1}), it was not significantly different from that on yellow flag days with rescues (0.23 ms^{-1}). The combined oscillatory and mean current of >1.0 ms^{-1} is significantly larger than for all red flag days and yellow flag days with rescues, suggesting that red flag days with rescues depend on inshore wave height and oscillatory currents.

WATER LEVEL

Dangerous rip currents are often linked to lower tidal elevations (Shepard et al., 1941; McKenzie, 1958; Short and Hogan, 1994; Lushine, 1991; Lascody, 1998; Engle et al., 2002). However, most drownings at Pensacola Beach occurred during periods of elevated water level, averaging 0.19 m above msl (Table 11.1). Similarly, rescues during yellow and red flag conditions are associated with water levels greater than msl, while green flag rescues are associated with water levels close to 0 m msl.

To examine the role of water level in controlling rip development, cross-shore variations in significant wave height are plotted for green, yellow, and red flag conditions in Figure 11.8. Wave height is unaffected by the outermost bar during green flag conditions (Figure 11.8a). There is slightly greater dissipation over the shoal with decreasing water levels, until the shoal is exposed as the water level approaches 0.3 m. As a consequence, maximum set-up at the shoreline is inversely related to water depth, leading to stronger rip currents at low water levels. In contrast, wave heights are attenuated by the outermost bar during yellow and red flag conditions, leading to an opposite relationship between maximum set-up and water depth. Increasing depths allow for larger waves inshore and greater dissipation across the shoal. This

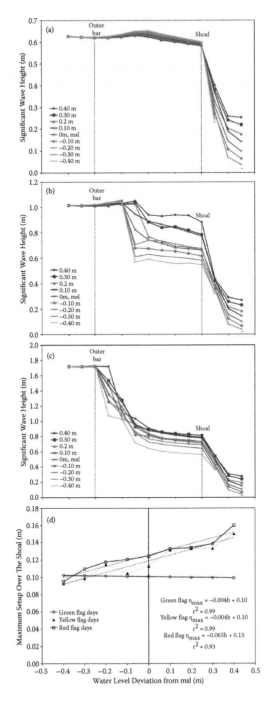

FIGURE 11.8 (*See color insert.*) Modeled cross-shore transformation of incident wave fields for (a) green, (b) yellow, and (c) red flag conditions with respect to inner- and outermost bars; (d) is variation in maximum set-up across the shoal with respect to water level.

does not necessarily lead to increased rip speeds as there is also an increase in wave heights in the rip channel under red flag conditions.

Tides at Pensacola Beach are diurnal; high tide occurs around 8 am at the onset of spring tides, 10 am at the height of spring tides, and 12 am at the transition to neap tides. To examine the daily variations in rip hazard in response to this consistent water level, water depths, wave heights, and predicted rip current speeds were ensemble-averaged by hour of day between March and September 2000 to 2008 (Figure 11.9). While the offshore significant wave height reached a late afternoon (4 pm) maximum due to developing sea breezes, the inshore significant wave height tended to reach maximum around 11 am in response to maximum sea level. Rip speed is predicted to reach a sustained maximum from 11 am to 1 pm in response to the greater water depths and larger inshore wave heights.

As the water levels increase and breaking over the outermost bar is reduced, more frequent wave sets come ashore, creating a flash rip effect (Bob West, personal communication). This suggests that in addition to the stronger mean current resulting from dissipation of larger inshore waves, additional pulsation of the current at infragravity frequencies (Sonu, 1972; Aagaard et al., 1997; Brander and Short, 2000; MacMahan et al., 2003; Reniers et al., 2004) cannot be addressed by the model. However, the times of above-average wave heights and rip speeds overlap significantly with heaviest beach use and account for 84% of all drownings.

DISCUSSION

The Florida panhandle is "… the worst [area] in the nation for beach drownings" (*Tuscaloosa News*, April 6, 2002). Santa Rosa Island was designated the "Drowning Capital" (*Lakeland Ledger*, April 3, 2002) after eleven drownings in 2001 (3 at Pensacola Beach). Between 2000 and 2008, twenty-five drownings occurred at Pensacola Beach. The drowning hot spots are situated between transverse ridges where the waves appear smaller and conditions safer. In addition, rip hot spots are coincident with the primary beach access points and resulting clusters of beach users at those locations.

Model results suggest that the average rip current speed on days with rescues ranges from 0.18 to 0.24 ms^{-1} and from 0.19 to 0.33 ms^{-1} for drowning events. In comparison, predicted current speeds for periods without rescues or drownings are on the order of 0.15 ms^{-1} regardless of the flag color flying that day. The days with drownings and rescues exhibit no statistically significant dependence on wind and wave forcing, but are significantly over-represented for water levels greater than msl and for tidal ranges exceeding the mean tidal range. An increase in rip speeds at high tide is contrary to previous studies linking rip current danger to lower tidal elevations (Shepard et al., 1941; McKenzie, 1958; Short and Hogan, 1994; Lushine, 1991; Lascody, 1998; Engle et al., 2002). Lower tides during storm waves increase wave dissipation and set-up across the bar and drive larger rip current velocities until the bar emerges and the currents are driven only by water trapped landward of the bar.

At Pensacola Beach, rip current speed is inversely related to water level during green flag conditions (Figure 11.8d), while low water levels during yellow and red flag conditions are associated with an increase in wave breaking and attenuation over

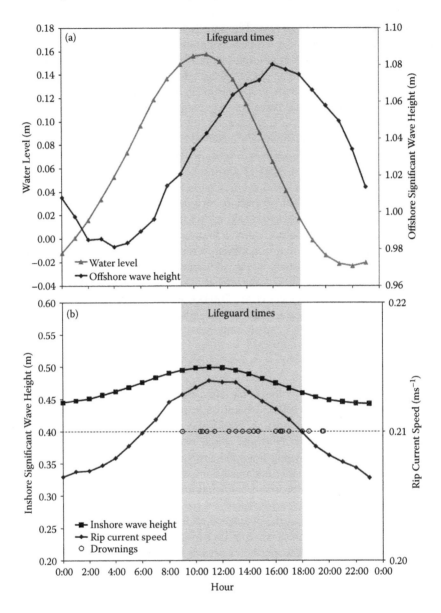

FIGURE 11.9 Ensemble-averaged hourly variation in (a) water level and offshore significant wave height and (b) inshore significant wave heights, rip current speeds, and drowning times.

the outermost bar. This limits the heights of inshore waves breaking on the shoal.
Therefore, the morphology and behavior of the innermost bar and its rip channels
are partly controlled by position and relative depth of the outermost bar (Ruessink
and Terwindt, 2000; Masselink, 2004; Houser and Greenwood, 2005; Castelle et al.,
2006; Castelle et al., 2010).

While drownings tend to occur on red flag days, lifeguards at Pensacola Beach regularly perform rescues, assists, and prevents on green and yellow days. This in part reflects the semi-permanent nature of the channels that allows rip current development under the right combination of waves and tides. On green flag days, the current begins to develop in late afternoons, with falling tide and an increase in sea breeze (Figure 11.9). The strength of the rip under green flag conditions is surprisingly close to conditions for both red and yellow flags (Table 11.4), which may reflect more water entering the channel from the side of the shoal and a less focused current (McKenzie, 1958; Sonu, 1972; Short, 2007).

Rip current speeds cited in this study do not include an infragravity component that can make a current stronger (Sonu, 1972; Aagaard et al., 1997; Brander and Short, 2000; MacMahan et al., 2003; Reniers et al., 2004). As noted by West (personal communication), "more frequent wave sets" create a strong "flash rip current effect" as the water level increases and breaking over the outermost bar is reduced. Houser and Barrett (2009, 2010) also noted a strong infragravity signal in the elevation of the swash zone during high tide on a yellow flag day. While the infragravity signal was dominant throughout high tide, it was at a maximum with the incoming and outgoing tides in response to the decrease in water level. MacMahan et al. (2006) suggested that the infragravity component of the current can be double that of the mean current, suggesting that the rip velocities at Pensacola Beach may reach 0.72 ms^{-1} under yellow and red flag conditions.

Bathers are also subject to relatively large waves and oscillatory currents within the rip channel. Specifically, the model predicts wave heights between 0.47 and 0.86 m within the channel, consistent with the observations of Sonu (1972) at a similar site. The larger waves within the channel are in part responses to shoaling through the deeper rip channel, but also wave–current interactions that, while weak (Froude number of 0.06 to 0.11 for drownings), nonetheless increase wave heights. The presence of larger, unbroken waves in the rip channel may limit further current development through opposing radiation stress (Yu and Slinn, 2003) that forces a counter torque that opposes the pressure gradient from shoal to rip (Haller et al., 2002; Haas et al., 2003). Nonetheless, a pressure gradient exists between the shoal and the rip channel since the waves in the rip channel break and set up closer to shore. A bather entering a rip channel would encounter rapid changes in depth and short but very strong oscillatory pulses (0.7 to 1.1 ms^{-1}), possibly leading to unstable and lost footing in depths of 1 to 1.5 m.

Strong tidal dependence of rips at Pensacola Beach and the tendency of high tides to occur at mid-day through the summer season mean that these dangerous currents reach maximum speed when most people are at the beach (Figure 11.9). This explains why rip currents at Pensacola Beach and elsewhere along Santa Rosa Island are responsible for so many drownings and rescues.

CONCLUSIONS

The primary access points at Pensacola Beach are located where semi-permanent rip channels exist, and model results indicate that rips can occur over a wide range of wave and tidal conditions. Rip strength is controlled by wave breaking over the outermost bar and reaches a maximum at high tide, which occurs mid-day for spring tides; red and yellow flags are flown at these times. Rip currents can still develop during green flag days; their strength increases during late afternoons in response to greater dissipation over the shoal caused by falling tide and onset of sea breezes.

The model predicts that the mean rip current velocity on rescue days is relatively weak (0.2 to 0.3 ms^{-1}), but this does not include pulsing flows resulting from infragravity waves. Direct measurements of rip speeds, locations, and timings are required to confirm the model results and estimate the relative importance of infragravity waves.

ACKNOWLEDGMENTS

This study was supported by in part by grants from the National Park Service (P5320060026) and Florida Sea Grant (R/C-S-50). Field assistance was provided by Tim Brunk and Jack Walker of Texas A&M University and by Tanya Gallagher, Fritz Langerfeld, and Klaus Meyer-Arendt from the University of West Florida. Bob West of the Santa Rosa Island Authority is also thanked for his support and contributions to this study.

REFERENCES

Aagaard, T., B. Greenwood, and J. Nielsen. 1997. Mean currents and sediment transport in a rip channel. *Marine Geol.,* 140: 25–45.

Bowman, D., D. Arad, D.S. Rosen et al. 1988. Flow characteristics along the rip current system under low energy conditions. *Marine Geol.,* 82: 149–167.

Brander, R.W. 1999. Field observations on the morphodynamic evolution of a low energy rip current system. *Marine Geol.,* 157: 199–217.

Brander, R.W., and A.D. Short. 2000. Morphodynamics of a large-scale rip current system at Muriwai Beach, New Zealand. *Marine Geol.,* 165: 27–39.

Brander, R.W., A.D. Short, P.D. Osborne et al. 1999. Field measurements of a large-scale rip current system. In *Coastal Sediments,* Kraus, N.C. and McDougal, W.G., Eds.

Castelle, B., P. Bonneton, and R. Butel. 2006. Modeling of crescentic pattern development of nearshore bars: Aquitanian Coast, France. *Compt. Rend. Geosci.,* 338: 795–801.

Castelle, B., B.G. Ruessink, P. Bonneton et al. 2010. Coupling mechanisms in double sandbar systems 1: Patterns and physical explanation. *Earth Surface Proc. Landforms,* 35: 476–486.

Engle, J., J. MacMahan, R.J. Thieke et al. 2002. Formulation of a rip current predictive index using rescue data. *Proc. 15th Annu. Conf. on Beach Preservation Technology,* Tait, L., Ed.

Fletemeyer, J. and S. Leatherman. 2010. Rip currents and beach safety education. *J. Coastal Res.,* 26: 1–3.

Gensini, V. and W. Ashley. 2010. An examination of rip current fatalities in the United States. *Natural Hazards,* 54: 159–175.

Greenwood, B. and P.R. Mittler. 1979. Structural indexes of sediment transport in a straight, wave-formed nearshore bar. *Marine Geol.,* 32: 191–203.

Haas, K.A., I.A. Svendsen, M.C. Haller et al. 2003. Quasi three-dimensional modeling of rip current systems. *J. Geophys. Res. Oceans,* 108: C7.

Haller, M.C., R.A. Dalrymple, and I.A. Svendsen. 1998. Experimental modeling of a rip current system. In *Ocean Wave Measurement and Analysis,* Edge, B.L. and Hemsley, J.M., Eds.

Haller, M.C., R.A. Dalrymple, and I.A. Svendsen. 2002. Experimental study of nearshore dynamics on a barred beach with rip channels. *J. Geophys. Res. Oceans,* 107: C6.

Houser, C. and G. Barrett. 2010. Divergent behavior of the swash zone in response to different foreshore slopes and nearshore states. *Marine Geology*, 271: 106–118.

Houser, C. and B. Greenwood. 2005. Profile response of a lacustrine multiple barred nearshore to a sequence of storm events. *Geomorphology*, 69: 118–137.

Lascody, R.L. 2009. East Central Florida rip current program. *Natl. Weather Dig.*, 22: 25–30.

Livingston, G. and K. Arthur. 2002. The economic impact of Pensacola Beach (unpublished report). Haas Center for Business Research and Economic Development, University of West Florida. Pensacola.

Lushine, J.B. 1991. A study of rip current drownings and related weather factors. *Natl. Weather Dig.*, 16: 13–19.

MacMahan, J., A.J.H.M. Reniers, E.B. Thornton et al. 2004. Infragravity rip current pulsations. *J. Geophys. Res. Oceans*, 109: C1.

MacMahan, J.H., E.B. Thornton, and A.J.H.M. Reniers. 2006. Rip current review. *Coastal Eng.*, 53: 191–208.

MacMahan, J.H., E.B. Thornton, A.J.H.M. Reniers et al. 2008. Low-energy rip currents associated with small bathymetric variations. *Marine Geol.*, 255: 156–164.

Masselink, G. 2004. Formation and evolution of multiple intertidal bars on macrotidal beaches: application of a morphodynamic model. *Coastal Eng.*, 51: 713–730.

McKenzie, P. 1958. Rip current systems. *Journal of Geology*, 66: 103–113.

Mei, C.C., and P.L.F. Liu. 1977. Effects of topography on circulation in and near surf zone: linear theory. *Est. Coastal Marine Sci.*, 5: 25–37.

Meyer-Arendt, K.J. 1990. Recreational business districts in Gulf of Mexico seaside resorts. *J. Cultural Geogr.*, 11: 39–55.

Morgan, D., J. Ozanne-Smith, and T. Triggs. 2009. Self-reported water and drowning risk exposure at surf beaches. *Austral. New Zeal. J. Public Health*, 33: 180–88.

Reniers, A.J.H.M., J.A. Roelvink, and E.B. Thornton. 2004. Morphodynamic modeling of an embayed beach under wave group forcing. *J. Geophys. Res. Oceans*, 109: C1.

Ruessink, B.G. and J.H.J. Terwindt. 2000. The behaviour of nearshore bars on the time scale of years: a conceptual model. *Marine Geol.*, 163: 289–302.

Shepard, F.P., K.O. Emery, and E.C. Lafond. 1941. Rip currents: a process of geological importance. *J. Geol.*, 49: 338–369.

Short, A.D. 1985. Rip current type, spacing and persistence, Narrabeen Beach, Australia. *Marine Geol.*, 65: 47–71.

Short, A.D. 2007. Australian Rip Systems—Friend or Foe. *Journal of Coastal Research*, SI50: 7–11.

Short, A.D., and C.L. Hogan. 1994. Rip currents and beach hazards: their impact on public safety and implications for coastal management. *J. Coastal Res.*, 12: 197–209.

Smith, J.A. and J.L. Largier. 1995. Observations of nearshore circulation: rip currents. *J. Geophys. Res. Oceans*, 100: 10967–10975.

Sonu, C.J. 1972. Field observation of nearshore circulation and meandering currents. *J. Geophys. Res.*, 77: 3232.

Thornton, E.B., J. MacMahan, and A.H. Sallenger. 2007. Rip currents, mega-cusps, and eroding dunes. *Marine Geol.*, 240: 151–167.

Turner, I.L., D. Whyte, B.G. Ruessink et al. 2007. Observations of rip spacing, persistence and mobility at a long, straight coastline. *Marine Geol.*, 236: 209–221.

Wright, L.D. and A.D. Short. 1984. Morphodynamic variability of surf zones and beaches: a synthesis. *Marine Geol.*, 56: 93–118.

Yu, J. and D.N. Slinn. 2003. Effects of wave–current interaction on rip currents. *J. Geophys. Res. Oceans*, 108: C3.

12 Rip Currents in the Great Lakes

An Unfortunate Truth

Guy Meadows, Heidi Purcell, David Guenther, Lorelle Meadows, Ronald E. Kinnunen, and Gene Clark

CONTENTS

INTRODUCTION

The five Great Lakes are not characteristic of inland lakes; they are large inland seas, with typical dimensions of over 500 km in length and 150 km in width. Strong and rapidly changing wind fields produce severe, locally generated seas with significant wave heights in excess of 7 m during extreme conditions. Many people associate rip currents only with oceanic coasts and are surprised to discover that they do indeed occur with unfortunate regularity along the coastlines of the Great Lakes (Figure 12.1).

The Great Lakes experienced twenty-five rip-related deaths from January to early September 2010. Between 1994 and 2007, the state of Michigan was in fourth place on the list of greatest number of rip current fatalities in the continental U. S. (Gensini and Ashley, 2009). Figure 12.2 displays rip locations and total numbers for locations documented by the National Weather Service (NWS) for 2002 through 2009 when the Great Lakes Basin experienced an average of seven rip fatalities a year.

The Great Lakes Basin encompasses over 17,500 km of coastline, almost 2,000 km more than the United States Atlantic and Pacific coasts combined and much of the coastline is accessible to tourists. Based on this vast length of coastline and the corresponding large numbers of visitors and residents, rip currents in the Great Lakes are phenomena that should not be overlooked.

FIGURE 12.1   Rip currents along Grand Sable. (Courtesy of Don Rolfson.)

FIGURE 12.2   Great Lakes rip current fatalities, 2002 through 2009, as recorded by National Weather Service.

The scientific community and general public have long recognized rip currents as hazards on oceanic coasts. Over 20,000 surf rescues (80% of the total) and 150 fatalities are attributable to rip currents each year at U.S. beaches (Lushine, 1991a and b). It is widely believed that better forecasting and greater public awareness can mitigate these coastal hazards. The NWS produces surf zone forecasts for the Atlantic, Gulf, and Pacific coasts, but did not begin issuing Great Lakes rip forecasts until 2006. Several NWS offices in the Great Lakes area are currently testing rip current advisory forecast methods based upon wind speed, fetch, duration, and other variables. However, NWS cites a need for better nearshore wave models to help describe the unique characteristics and physical dynamics of the enclosed Great Lakes Basin.

Although strong cross-shore currents have been linked to Great Lakes drownings, little quantitative data exist to determine the hydrodynamic conditions at the times of the incidents. Rip currents are transient in nature, requiring immediate response to capture what amounts to "perishable" data. This chapter focuses on hypothesis-driven, nearshore dynamics to explain the generation of rip currents in the Great Lakes. These nearshore dynamics differ from those of oceanic coasts in several important ways.

RIP CURRENT RESEARCH

The first scientific observations of rip current circulations were made by Shepard et al. (1941) off the coast of La Jolla, California. For this site, Shepard and Inman (1950) determined that wave refraction over the offshore topography (i.e., submarine canyons) created strong longshore wave height gradients that led to rip current generation. Other mechanisms for rip current generation were identified through later studies: longshore wave height variation due to standing edge waves (Bowen, 1969; Bowen and Inman, 1969); intersecting wave trains of identical frequency leading to longshore variations in water level (Dalrymple, 1975); nearshore bathymetric variations (Haller et al., 2002); and coupled hydrodynamic–morphodynamic systems with variable longshore bathymetry and hydrodynamics (Hino, 1974).

Scientific observations of rip currents have been hampered by the difficulty in deploying instrumentation in this harsh surf environment. Despite limited observation sets, detailed examination of currents within rip channels commonly demonstrates pulsations on relatively long time scales; several explanations exist for this phenomenon. Rip pulsations were shown to be related to infragravity motions on the order of 0.004 to 0.04 Hz (Sonu, 1972; Suhayda, 1974; Wood and Meadows, 1975; Guza and Thornton, 1985; MacMahan et al., 2004a). Also, mass transport and wave set-up produced by larger waves within a wave group can cause significant "puddling" of water within a surf zone that returns to the sea via rip channels (Munk, 1949; Shepard and Inman, 1950). Additionally, longer period motions (≥15 min) of non-gravity waves associated with rip shear instabilities were cited as driving these pulsations (Smith and Largier, 1995; Brander and Short, 2001; Haller and Dalrymple, 2001). Finally, wave group-induced vortices may also contribute to rip current pulsations (Reniers et al., 2004; MacMahan et al., 2004b).

Wood and Meadows (1975) made the first field measurements in the Great Lakes of unsteadiness in longshore currents and observed long period wave-induced motions in the 30- to 90-sec period ranges. The fluctuations were persistent across the active surf zone and throughout the water column. In 1982, the spatial pattern in Lake Michigan of these wave motions was shown to be observable from space via synthetic aperture radar (Meadows et al., 1982).

A secondary research focus has been morphological control of rips through laboratory, field, and numerical studies (Van Enckevort et al., 2004; MacMahan et al., 2006). In some cases, rip channels are established through interactions of waves with offshore topographic features, such as submarine canyons (Long and Ozkan-Haller, 2005). On many barred beaches, rip channels are expressed as quasi-regularly spaced channels flanked by crescentic bars, and research has focused on channel spacing as related to incident wave conditions or morphodynamic dimensions. Unfortunately, this research focus has not been fruitful (Huntley and Short, 1992). Time-lapsed video images of the surf zone have been successfully used for data retrieval (Van Enckevort et al., 2004; Holman et al., 2006), but analysis failed to support any causal relationships. Numerical investigations also fall short in providing robust estimates of rip channel spacing (Damgaard et al., 2002). Calvete et al. (2007) showed through numerical sensitivity analysis that the lack of predictability of rip channel development and spacing is an inherent property of coastal systems and is related to the pre-existing bathymetry of the nearshore zone.

Understanding the development of Great Lakes rips requires acquisition of time-sensitive ("perishable") data before, during, and after an event to determine whether nearshore morphodynamic features play a role in rip current generation. For example, two photographs were taken seconds apart as an aircraft proceeded along the Lake Michigan shoreline (Figure 12.3). The incident wave field is nearly identical in both areas, but one exhibits a long linear, three-bar system while the other shows a series of rips.

Rip currents have been observed to vary with tidal elevation on oceanic coasts. Although the *rip tide* misnomer infers a relationship with tidal currents, in actuality, the tidal influence is due to changes in water level accompanying tidal cycles. Decreases in tidal elevation tend to increase rip speed (Sonu, 1972; Brander and

FIGURE 12.3   Coastal aerial photographs of two adjacent sections of Lake Michigan shoreline along Big Sable Point. Left: long linear, three-bar system. Right: complex rip channel system appearing a few seconds later as the aircraft progressed.

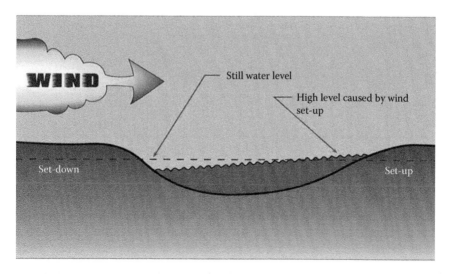

FIGURE 12.4   Seiche caused by water buildup due to wind stress or pressure gradient in an enclosed basin. (*Source:* Modified from Great Lakes Atlas, http://www.epa.gov/glnpo/atlas/index.html)

Short, 2001; MacMahan et al., 2005). Although astronomical tides are minimal in the Great Lakes, seiches are similar in magnitude and duration to ocean tides.

Seiches result from consistent wind blowing over an enclosed basin, resulting in a mass transport of surface water toward the downwind coast and a resultant increase in water level (Figure 12.4). When the wind ultimately decreases or shifts, this build-up of water is free to flow back, resulting in a complex rotational wave within the basin, not dissimilar from a tidal bulge along an ocean coast. For example, seiches (wind tides) both contribute excess water to the nearshore zone and can significantly change wave elevations over relatively short periods.

Large variations in water level are common in the Great Lakes on a range of time scales not seen on oceanic coasts. Water level variations on the order of 0.3 m occur seasonally, with variations as large as 0.5 m annually and extremes approaching 2 m on decadal time scales. These longer term fluctuations change beach and nearshore slopes, cause migration of semi-permanent sand bars, and alter nearshore sediment supplies (Meadows et. al., 1997). Since the waters filling the Great Lakes are primarily the results of evaporation and precipitation, even the incident wave climate varies with the frequencies and intensities of storms on decadal time scales. Figure 12.5 depicts a representative section of the long-term water level record for Lakes Michigan and Huron from 1918 through 2002. This historical record clearly demonstrates the multiple time scales of Great Lakes water elevation variations.

Wind-generated waves in enclosed basins are by definition locally generated (resulting from cyclones present over the basins). Such seas are complex, unsorted, steep, widely distributed in direction, and often accompanied by strong longshore wind components. All these factors combine to contribute to strong and difficult-to-predict nearshore circulations along the Great Lake coastlines.

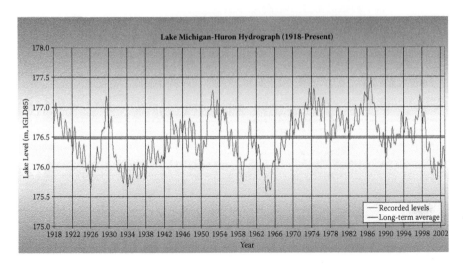

FIGURE 12.5   Representative section of Lake Michigan–Huron water levels, 1918 through 2002, as recorded by NOAA.

ENCLOSED BASIN NEARSHORE DYNAMICS

The Great Lakes are similar to oceans in many ways, but pointed differences make rip generation in an enclosed basin somewhat unique. The changing water levels (on many time scales) of the Great Lakes affect the locations and movements of nearshore sand bars that in turn influence the locations and formations of rip current channels. It is known that increased wind and higher wave energy precede an increase in long-term water levels (Meadows et al., 1997), which may set a "bathymetric stage" for increased rip frequency. In general, these conditions lead to a steeper, more mobile nearshore zone. Seiches—common to all the Great Lakes—are the most extreme in Lake Erie—a relatively shallow basin, oriented east to west into the prevailing wind with recorded elevation differences between Toledo, Ohio and Buffalo, New York of 4.9 m. These seiches oscillate in the longitudinal mode with a period of approximately 16 hr.

Seiching contributes excess water to the nearshore zone and can significantly change water elevation over relatively short periods. For example, the series of drownings along the southeast shoreline of Lake Michigan on July 4, 2003 was associated with a moderate to strong seiche (Guenther, 2003). During this single event, seven rip-related drownings were reported within a 3-hr period along a 3-mile section of state park beach. This area is ringed by National Ocean Service (NOS) water level gauges (Figure 12.6 and Figure 12.7) indicating that both transverse and longitudinal seiches were present in southern Lake Michigan at the time of the fatalities. This seiche was caused by the passage of an intense squall line and corresponding wind shift.

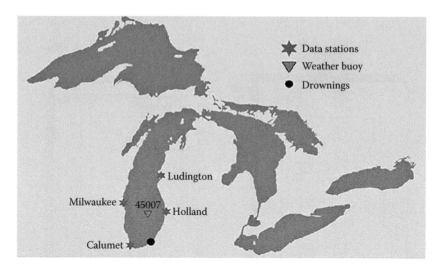

FIGURE 12.6 July 4, 2003 drowning event location in Lake Michigan, showing surrounding data stations.

Four men drowned in 2002 at Nickel Plate Beach in Huron, Ohio because of a strong rip current. The NOS records again indicated a seiche spanning the times of the drownings. The two records are from NOS stations at Marblehead, Ohio and Sturgeon Point, New York (Figure 12.8).

Dimensions of the Great Lakes are comparable in spatial scale to the atmospheric low and high pressure systems that traverse this region. This produces an incident wave climate of locally generated, fetch-limited seas that are complex, unsorted, and widely distributed in direction. This is in stark contrast to the conventional wisdom of rip current generation wherein a well-organized wave field is required to produce a correspondingly well-organized nearshore circulation. These various factors make rip current prediction difficult. Strong air and sea temperature differences develop in the Great Lakes during the fall. Cold outbreaks of Canadian polar air drive large and rapid wave growth, resulting in the infamous storms known as "the Gales of November." Notable Great Lakes shipwrecks that have occurred during "the Gales" include:

- November 10, 1975—*M/V Edmund Fitzgerald* (entire 29-member crew lost)
- November 18, 1958—*M/V Carl D. Bradley* (31 crew members of 33 lost)
- November 7–11, 1913—Great Storm of 1913 (251 people lost from 12 ships)

The combination of warm water temperatures coupled with strong wind outbreaks in November produce very dangerous Great Lakes wave conditions. Although still present, this threat decreases in the late summer months (August and September) when the beaches are still used for bathing. Under these unstable atmospheric conditions, wave growth is alarmingly fast, often catching bathers, surfers, and first

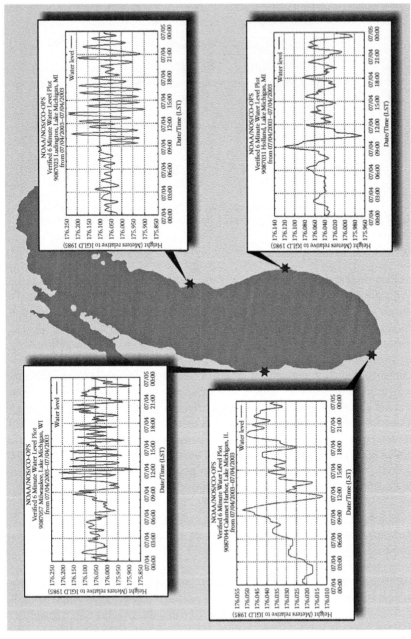

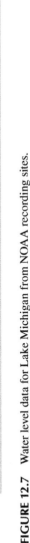

FIGURE 12.7   Water level data for Lake Michigan from NOAA recording sites.

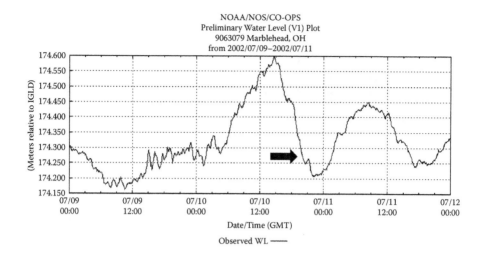

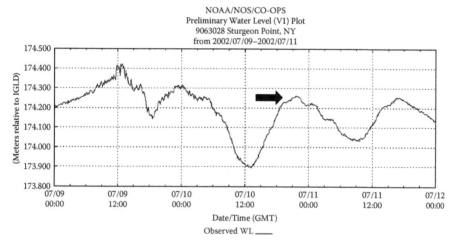

FIGURE 12.8 NOS water level records for Lake Erie for July 9–11, 2002. The arrow points to the water level at the time of the rip current drownings.

responders off guard. Figure 12.9 and Figure 12.10 from NOAA provide examples of such rapid and severe wave growth.

GREAT LAKES RIP FORECASTS

The dynamics of the Great Lakes enclosed basin and the locally generated seas provide many difficulties for forecasting rip currents. In 2006, the NWS Great Lakes offices began issuing surf zone forecasts that included rip risk information. Forecasting Great Lakes rips is a 3-step process. First, forecasters collect data on the observed conditions. The Great Lakes are surrounded by a comprehensive network of observing platforms where ground truth information can be obtained to verify forecasts as needed. The network includes:

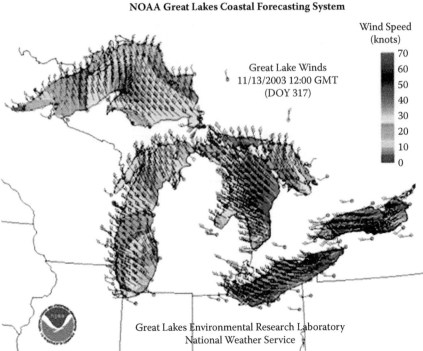

NOAA Great Lakes Coastal Forecasting System

Great Lake Winds
11/13/2003 12:00 GMT
(DOY 317)

Wind Speed
(knots)

70
60
50
40
30
20
10
0

Great Lakes Environmental Research Laboratory
National Weather Service

FIGURE 12.9   (*See color insert.*) Strong, late season outbreak of Canadian air over the Great Lakes Basin.

- **Buoys**—The U.S. National Data Buoy Center (NDBC), Environment Canada, and the recently added Great Lakes Observing System (GLOS) coastal monitoring network with directional wave capabilities all maintain buoys that provide wave height, wind speed and direction, and wave period data. The GLOS buoys and some NDBC buoys also provide wave directional information. NDBC and Environment Canada buoys are located in mid-lakes; GLOS buoys are sited along the coasts; all provide valuable real-time data.
- **NOS gauges**—NOS maintains a network of 54 tide gauges around the Great Lakes that measure lake levels at 60-min intervals and provide wind, air temperature, relative humidity, and barometric pressure data that are particularly helpful in monitoring for seiches.
- **Doppler radar**—The NWS and Environment Canada blanket the Great Lakes with Doppler radar coverage, providing real-time precipitation and wind data aloft.
- **Ship observations**—Numerous boats and ships voluntarily submit observations at regular intervals as they traverse the lakes, providing temperature, wind, and wave information.

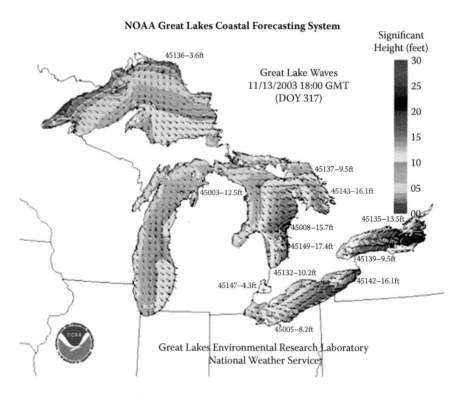

NOAA Great Lakes Coastal Forecasting System

Great Lake Waves
11/13/2003 18:00 GMT
(DOY 317)

Significant
Height (feet)

Great Lakes Environmental Research Laboratory
National Weather Service

FIGURE 12.10 (*See color insert.*) Resulting Great Lakes wave field (significant wave height in feet).

- **Nearshore automated observation platforms**—Numerous automated aviation and marine observing stations have been installed within a few kilometers of the lakeshore.
- **Satellite data**—Satellites provide sky cover, ice condition, and temperature profiles over the lakes.
- **Rawinsonde stations**—Several rawinsonde stations are in close proximity to the Great Lakes and provide wind and temperature profiles of air masses over the lakes. They take readings twice a day.

Forecasts of wind speed and directions for the Great Lakes are made by operational forecasters at all NWS forecast offices based on the voluminous data collected. The wind forecasts are then used to predict the wave characteristics on the lakes. The rip current risk for each area is then determined by applying the Great Lakes Rip Current Checklist (GLRCC). See Figure 12.11. The GLRCC was developed for the Great Lakes by modifying the East Central Florida Lushine Rip Current Scale (ECFL LURCS; Engle et al., 2002) and the Lushine Rip Current Scale (LURCS; Lushine, 1991) using parameters unique to the Great Lakes. The checklist works well for the long fetch parts with sandy beaches. For breakwaters, groins, and

Great Lakes Rip Current Checklist

I. Wave Factors	Degrees from Normal	
Height	**0 to 30**	**30 to 70**
1 ft	0.5	0.0
2 ft	1.0	0.5
3–4 ft	2.0	1.0
5–7 ft	3.0	1.5
8–10 ft	4.0	2.0
>10 ft	5.0	3.0

Wave factor _____

II. Wind Speed	
Speed	**points**
0–5 kt	0.5
6–8 kt	1.0
9–11 kt	1.5
12–14 kt	2.0
15–17 kt	3.0
18–20 kt	4.0
>20 kt	5.0

Wind factor _____

III. Wave Period in Seconds	
5–6 sec	0.5
7–8 sec	1.0
9–10 sec	2.0
>10 sec	3.0

Period factor _____

IV. Lake level vs. normal		
above	>3 in	–0.5
	normal	0.0
below	–6 to 11 in	1.0
	> –11 in	2.0

Lake level factor _____

Total _____

low risk < 4   moderate risk = 4 to 7   high risk > 7

FIGURE 12.11   Great Lakes rip current checklist.

jetties, forecasters assume that rips will occur where longshore currents are diverted offshore by these shore-perpendicular structures. Waves with heights of 0.6 m or more are predicted to produce moderate to high risks of rip currents. Forecasters adjust their criteria based on feedback from local users at specific beaches.

Rip forecasts are disseminated by the Advanced Weather Interactive Processing System (AWIPS; Glahn and Ruth, 2003) that routes data to the media and on the Internet. Surf forecasts are produced by the six offices throughout the Great Lakes that are most susceptible to rip currents (Figure 12.12).

CONCLUSIONS

The Great Lakes have both the size and meteorological conditions necessary to develop large waves and rip currents that pose significant threats to beachgoers. Most rip research has been conducted in ocean environments, and these dangerous currents should be more thoroughly investigated in the Great Lakes. Field data must be collected during rip events to determine whether morphodynamics play a role in their generation. Seiches have been recorded in conjunction with a number of rip current drownings, and this relationship must be quantified. Large water level variations in the Great Lakes influence bar morphology and location, and also affect rip formation.

```
SURFZONE FORECAST
NATIONAL WEATHER SRVICE DULUTH MN
450 AM CDT TUE SEP 28 2010


MNZ037-WIZ001-282315-
CARLTON/SOUTHERN ST. LOUIS-DOUGLAS-
INCLUDING THE BEACHES OF...DULUTH...CLOQUET...SUPERIOR
450 AM CDI TUE SEP 28 2010


.TODAY...
SKY/WEATHER........MOSTLY CLOUDY (80-90 PERCENT) UNTIL 2 PM...THEN
                   PARTLY CLOUDY (50-60 PERCENT).
MAX TEMPERATURE....53-58.
BEACH WINDS.......EAST WINDS AROUND 10 MPH.
SURF..............1 TO 3 FEET.
RIP CURRENT RISK...MODERATE. A MODERATE RISK OF RIP CURRENTS MEANS
                   WIND AND OR WAVE CONDITIONS SUPPORT STRONGER OR
                   MORE FREQUENT RIP CURRENTS. ALWAYS HAVE A
                   FLOTATION DEVICE WITH YOU IN THE WATR.
```

FIGURE 12.12 Example of surf zone forecast

Rip current forecasts are dependent upon such data collection and analysis and are essential to ensure the Great Lakes beaches are safe for residents and tourists alike.

REFERENCES

Bowen, A.J. 1969. Rip currents 1. Theoretical investigations. *J. Geophys. Res.*, 74: 5467–5478.

Bowen, A.J. and D.L. Inman. 1969. Rip currents 2. Laboratory and field observations. *J. Geophys. Res.*, 74: 5479–5490.

Brander, R.W. and A.D. Short. 2001. Morphydynamics of a large-scale rip current system at Muriwai Beach, New Zealand. *Marine Geol.*, 165: 27–39.

Calvete, D., G. Coco, A. Falques et al. 2007. Unpredictability in rip channel systems. *Geophys. Res. Lett.*, 34: L05605, doi:10.1029/2006GL028162.

Dalrymple, R.A. 1975. A mechanism for rip current generation on an open coast. *J. Geophys. Res.*, 80: 3485–3487.

Damgaard, J., N. Dodd, L. Hall et al. 2002. Morphydynamic modeling of rip channel growth. *Coastal Eng.*, 45: 199–221.

Engle, J., J. MacMahan, R.J. Thieke et al. 2002. Formulation of a rip current predictive index using rescue data. *Proc. Natl. Conf. on Beach Preservation Technology.* Florida Shore and Beach Preservations Association, Biloxi, MS.

Gensini, V. and W. Ashley. 2009. An examination of rip current fatalities in the United States. *Natural Haz.*, 54: 159–175.

Glahn, H.R. and D.P. Ruth. 2003. New digital forecast database of the National Weather Service. *Bull. Amer. Meteorol. Soc.*, 84: 195–202.

Guenther, D. 2003. Rip current case study 3, 4 July 2003. *Marquette Michigan National Weather Service Office Report.*

Guza, R.T. and E.B. Thornton. 1985. Observations of surf beat. *J. Geophys. Res.*, 90: 3161–3171.

Haller, M.C. and R.A. Dalrymple. 2001. Rip current instabilities. *J. Fluid Mech.*, 433: 161–192.

Haller, M.C., R.A. Dalrymple, and I.A. Svendsen. 2002. Experimental study of near-shore dynamics on a barred beach with rip channels. *J. Geophys. Res.*, 107: 3061, doi:10.1029/2001JC000955.

Hino, M. 1974. Theory on formation of rip-current and cuspidal coast. *Coastal Eng.*, 901–919.

Holman, R.A., G. Symonds, E.B. Thornton et al. 2006. Rip spacing and persistence on an embayed beach. *J. Geophys. Res.*, 111: C01006, doi:10.1029/2005/JC002965.

Huntley, D.A. and A.D. Short. 1992. On the spacing between observed rip currents, *Coastal Eng.*, 17(3–4): 211–225.

Lascody, R.L. 1998. East Central Florida rip current program. *Natl. Weather Dig.*, 222: 25–30.

Long, J.W. and H.T. Ozkan-Haller. 2005. Offshore controls on nearshore rip currents. *J. Geophys. Res.*, 110: C12007, doi:10.1029/2005JC003018.

Lushine, J.B. 1991a. A study of rip current drowning and related weather factors. *Natl. Weather Dig.*, 16: 13–19.

Lushine, J.B. March 2005b. Rip current formation and forecasting. Presented at Beach Safety Educational Workshop, St. Petersburg, FL

MacMahan, J.H., E.B. Thornton, and J.H.M. Reniers. 2006. Rip current review. *Coastal Eng.*, 53: 191–208.

MacMahan, J.H., J.H.M. Reniers, E.B. Thornton et al. 2004a. Infragravity rip current pulsations. *J. Geophys. Res.*, 109: C07004, doi:10.1029/2003JC002083.

MacMahan, J.H., J.H.M. Reniers, E.B. Thornton et al. 2004b. Surf zone eddies coupled with rip current morphology. *J. Geophys. Res.*, 109: C01003, doi:10.1029/2003JC002068.

MacMahan, J.H., E.B. Thorton, T.P. Stanton, and A.J.H.M. Reiniers. 2005. RIPEX: Observations of a rip current system, *Marine Geol.*, 218(1–4): 113–114.

Meadows, G.A., L.A. Meadows, W.L. Wood et al. 1997. Relationship between Great Lakes water levels, wave energies and shoreline damage. *Bull. Amer. Meteorol. Soc.*, 78: 675–683.

Meadows, G.A., R.A. Shuchman and J.D. Lyden. 1982. Analysis of remotely sensed long-period wave motions. *J. Geophys. Res.*, C8: 5731–5740.

Munk, W.H. 1949. Surf beats. *EOS Trans. Amer. Geophys. Union*, 306: 849–854.

NOAA. 2005. *New Priorities for the 22st Century: NOAA's Strategic Plan Updated for FY2006–FY2011*. U.S. Department of Commerce, Washington, pp. 1–24.

Reniers, J.H.M., J.A. Roelvink, and E.B. Thornton. 2004. Morphydynamic modeling of an embayed beach under wave group forcing. *J. Geophys. Res.*, 109: C01030, doi:10.1029/2002JC001586.

Shepard, F.P. and D.L. Inman. 1950. Nearshore water circulation related to bottom topography and wave refraction. *EOS Trans. Amer. Geophys. Union*, 312: 196–212.

Shepard, F.P., K.O. Emery, and E.C. La Fond. 1941. Rip currents: a process of geological importance. *J. Geol.*, 49: 337–369.

Smith, J.A. and J.L. Largier. 1995. Observations of nearshore circulation: rip currents. *J. Geophys. Res.*, 100: 10967–10975.

Sonu, C.J. 1972. Field observation of nearshore circulation and meandering currents. *J. Geophys. Res.*, 77: 3232–3247.

Suhayda, J.N. 1974. Standing waves on beaches, *J. Geophys. Res.*, 79: 3065–3071.

Van Enckevort, I.M.J., B.G. Reussink, G. Coco et al. 2004. Observations of nearshore crescentic sandbars. *J. Geophys. Res.*, 109: C06028, doi:10.1029/2003JC002214.

Wood, W.L. and G.A. Meadows. 1975. Unsteadiness in longshore currents. *Geophys. Res. Lett.*, 211: 503–505.

APPENDIX: CASE STUDY OF RIP CURRENT DROWNINGS AT HURON, OHIO

On July 10, 2002, four men between the ages of 18 to 34 drowned in a strong rip current while attempting to rescue a young woman from the waters off Nickel Plate Beach on the northeast side of Huron, Ohio at the west end of Lake Erie (Figure A.1).

FIGURE A.1 Nickel Plate Beach, Huron, Ohio.

At 18:30 UTC, a woman swimming at a sandbar screamed for help. She was in about 1.5 m depth approximately 23 m from shore. The rip current pulled her northeast, away from the shoreline. The men entered the water in an attempt to rescue her and drowned. An off-duty firefighter eventually reached the young woman and was able to rescue her with the help of others from the fire department. Waves, reported to be 1.2- to 1.8-m high, were strong enough to damage a boat used by the fire department during the rescue. Prior to the incident, red flags were posted by the city of Huron to warn swimmers to stay out of the water. Normally, a lifeguard is on duty; however, after red flags are posted, lifeguards are permitted to leave the beach.

NDBC Buoy 45005, located in western Lake Erie, 32 km north of the city of Huron, recorded winds from 040 to 050 degrees (northeast) at approximately 11 ms^{-1} for 5 hr prior to the incident. The winds were out of the north and northeast up to 21 hr before the incident. During the same period, the buoy reported waves 1.5 to 2.1 m high and a water temperature of 23 degrees C. The Automated Surface Observing System (ASOS) station at Cleveland, approximately 80 km east of Huron, reported winds from 030 degrees at 7.5 ms at 15:00 UTC. By 18:00 UTC, the winds increased to 12.5 ms^{-1} from 040 degrees. At Mansfield, Ohio, approximately 110 km south of Huron, weather observing equipment reported winds between 030 and 040 degrees for 4 hr prior to the accident and varying between 27 and 32 ms^{-1}. At 14:00 UTC, the winds diminished rapidly to 5.3 ms^{-1}. The Marblehead water level data gauge, about 25 km northwest of Nickel Plate Beach, indicated the lake reached a peak of 174.60 m mean surface level (msl) at 15:00 UTC and dropped to 174.44 m by 19:00 UTC and to 174.08 m by 22:00 UTC. After 22:00 UTC, the lake level started increasing again to peak out at 174.45 m at 08:00 UTC.

Nickel Plate Beach is oriented northwest to southeast (Figure A.1) and is thus highly exposed to northeast winds. Winds measured at Buoy 45005 and at the Cleveland airport appear to be most representative of the winds over Nickel Plate Beach at the time of the incident. Studies by Lushine (1991), Lascody (1998), and others show that a wind normal to shore would have a much higher probability of producing a dangerous rip current than one from any other direction. In addition to the high waves, reports from people at the beach noted a strong undertow. Surface water temperatures around 23 degrees C would indicate that hypothermia was not an issue.

Gary Packan, director of the Parks and Recreation Department for Huron, stated that the city had no specific criterion for posting red flag alerts and posted them when the waves looked high. The decision to post alerts is made by the city manager, parks director, and/or fire chief with input from local personnel.

The falling lake level indicated by the Marblehead gauge seems to indicate that a seiche occurred and further enhanced the strength of the rip current. The breakwaters north of Nickel Plate Beach may also have caused channeling and wave reflection, enhancing the subsequent rips.

Applying data from the buoy at the west end of Lake Erie to the rip current checklist indicates a wave factor of 3.0; wind speed factor based on the Cleveland airport ASOS station was 5.0. Estimating wave period based on wave heights at the buoy would yield a period of 4 sec and a factor of 0. According to the U.S. Army Corps of Engineers (2010), the average level for Lake Erie for July 10, 2002 was 174.24 m, and the Marblehead gauge showed a lake level of 174.44 m. Even though a seiche was occurring and the lake level at Marblehead dropped 0.16 m, the lake level at the time of the drownings was still above the monthly average. Therefore, the lake level factor would be −0.5. Adding these values together yielded a rip current risk of 7.5, which is classified as high.

13 Beach Safety Management in Brazil

Lauro J. Calliari, Antonio Henrique da F. Klein,
Miguel da G. Albuquerque, and Onir Mocellin

CONTENTS

INTRODUCTION

Coastal zones can be viewed as zones of convergence of economic and cultural factors. Three-quarters of the global population is predicted to live within 60 km of coastlines by 2020 (Povh, 2000). The main reasons for this increase are linked to the ease of construction and development of coastal tourism (Goya, 2009).

Scientific studies of oceanic beaches usually concern coastline stability and environmental quality. In terms of coastal management, beach safety is often a secondary consideration in most countries. However, surf beaches present significant hazards, particularly to users who have insufficient aquatic skills. Beach cities frequented by tourists must pay more attention to this problem.

Beach characteristics directly impact potential hazards and public safety. Short and Hogan (1993) classified beach hazards as permanent and non-permanent. Permanent hazards include promontories, reefs, rocky outcrops on beach faces, coastal structures such as jetties, breakwaters, platforms, shipwrecks, and other objects present in the surf zone. Inlet channels and water depth are also classified as permanent hazards, especially for poor swimmers. Non-permanent hazards display highly variable temporal scales and are mainly represented by wave breaker type and height, surf zone circulation, and beach morphologies such as numbers of bars and troughs, bar characteristics, and transverse channels. Beaches have been highlighted in the scientific literature due to the high numbers of bathing accidents and coastal management issues. According to the World Health Organization (WHO, 2004), more than 400,000 fatalities have been recorded annually at oceanic beaches (Table 13.1).

TABLE 13.1

Ocean Beach Drownings Worldwide

	World	RAF	RAM	RML	REU	RSEA	RPO
Males	281,717	67,654	20,181	20,712	30,322	55,258	87,600
Females	127,554	23,311	4,408	6,904	7,196	36,520	49,216
Total	409,272	90,965	24,589	27,616	35,518	91,778	136,816
Percent	100	22.3	6.0	6.8	9.1	22.4	33.4

Source: World Health Organization, 2004. Modified from Souza, P.H. 2005 Especialização em Planejamento e Controle em Segurança Pública. Universidade Federal do Paraná. Monografia de Conclusão de Curso.

Note: RAF = Africa region. RAM = Americas region. RML = Eastern Mediterranean region. REU = Europe region. RSEA = Southeast Asia region. RPO = West Pacific region.

AQUATIC RESCUE AND BEACH SAFETY PROGRAMS

Early in the twentieth century, Rio de Janeiro beaches started operating for leisure and tourism activities because they attracted year-round vacationers. Unfortunately, the beaches were hazardous because of high breaking waves and strong currents. Rio de Janeiro had one of highest drowning rates in Brazil. Commodore Wilbert E. Longfellow founded the American Red Cross Rescue Service in Rio de Janeiro—the first of its kind in Brazil.

The objective of the service was to organize and train volunteer lifeguards who would work at health stations. In 1939, the health station at Copacabana Beach was renamed for Ismael Gusmão as a tribute to its organizer. At that time, 18 fixed towers were built along the Rio de Janeiro coastline. A total of 120 lifeguards utilized motorized boats, ambulances, and resuscitating equipment. Bathing was restricted to areas in front of the towers and for specified hours. Rapid demographic growth rate and intense immigration to the city along with improvements in the standard of living in the 1950s increased beach use; this led to the creation of a rescue and life saving service dedicated to aquatic safety.

The Maritime Rescue Corps (Salvamar) was created in 1963; skilled and experienced people were recruited to handle aquatic rescues. In 1967, the Instruction Center for Salvage and Lifeguard Training was formed. Only 27 of 60 beaches were patrolled by the service which included 40 towers and 200 lifeguards. In 1967, the corps achieved 4,032 beach rescues with only 17 fatalities (Szpilman, 2004). The Maritime Rescue Corps became part of the Firefighters Brigade in 1984. Special efforts to train lifeguards greatly diminished the number of fatalities (Souza, 2005).

The state of Paraná had no rescue service before the1920s when beaches started to be used by bathers; life saving was usually performed by fishermen (Souza, 2005). In the mid 1950s, six civilian lifeguards were hired to work on Caiobá and Matinhos Beaches. Later this activity was transferred to the Firefighters Brigade, and the first specially trained lifeguards came from Rio de Janeiro. In the 1960s, water safety was implemented in other coastal counties under supervision of the Firefighters Brigade (Souza, 2005).

Educational programs were developed in 1990 under the influence of the Australian School of Coastal Geomorphology to reduce drowning accidents at oceanic beaches. The program resulted from a joint effort between Brazilian universities and the Firefighters Brigade. The University of Vale do Itajaí implemented the beach safety program along the Santa Catarina coast. This Brazilian initiative followed the same approach as the Australian beach safety programs (Hoefel and Klein, 1998; Pereira and Calliari, 2005; Albuquerque et al., 2010).

Interest in beach safety programs increased because drowning was the third most frequent cause of death in Brazil: 7,500 deaths annually in fresh and salt water and about 1.3 million surf rescues (Klein et al., 2003). In Paraná state, Angelotti (2004) found that 50% of the beach users had no swimming skills; the number of annual rescues recorded for this relatively short coast is impressive (Table 13.2). In addition, the study showed that 33% of the bathers swam at unpatrolled areas and 79% never consulted lifeguards about surf zone conditions.

In northeast Brazil, the development of coastal tourism has increased beach usage and the need to reduce the number of bathing accidents. Table 13.3 shows the number of bathing fatalities in the northeast coastal states from 2000 to 2008. In the southeast Brazilian states of Espirito Santo, Rio de Janeiro and São Paulo, beach safety studies indicate that 30% of bathers have no swimming skills. The principal causes of fatal and non-fatal accidents were rip currents.

TABLE 13.2
Annual and Daily Rescues on Coast of Paraná State (Summer Seasons)

	1998	1999	2000	2001	2002	2003	2004	2005
Annual rescues	1,354	996	1,354	1,394	1,429	1,120	1,377	1,656
Daily records	18.8	16.1	15.7	20.8	24.6	13.8	19.1	23.3

Source: From Souza, P.H. 2005. Especialização em Planejamento e Controle em Segurança Pública. Universidade Federal do Paraná. Monografia de Conclusão de Curso.

TABLE 13.3
Drowning Deaths in Northeast Brazil from 2000 to 2008

Bahia	3,700
Ceará	2,507
Pernambuco	2,133
Alagoas	1,008
Paraíba	952
Maranhão	909

Source: World Health Organization, 2009.

SOUTHERN BRAZIL

Studies by Hoefel and Klein (1998) focused on beach safety problems during the summers of 1995 and 1996 along fifteen microtidal beaches located in the central and northern parts of Santa Catarina state; 1,168 bathing rescues and seven deaths were documented. Most of the accidents (85%) were associated with rip currents despite the warning signs. The victims ranged in age from15 to 30 years old, and 50% were tourists from other states. The numbers of drownings related more to the intensity of beach usage and bather attitude than to the physical characteristics of the beaches (although semi-protected beaches with rocky headlands present greater risks because of persistent rip currents). Similar data obtained from 1995 and 2001 for 56 patrolled beaches along 532 km of coast suggest that two main factors favor accidents: people with inadequate understanding of surf beaches and rip currents that represent the main natural risks for bathers.

Analysis of more than 13,000 rescues during ten summer seasons, information about bathers, surf zone conditions, and identification of risky areas at twenty beaches in Santa Catarina indicated that the number of accidents directly related to beach exposure to swells from the southeast and east, breaker height, surf zone width, rip current occurrence, number of bathers, and easy beach access (Mocellin, 2006). Sixty-two percent of drowning victims were males and 38% females. Most victims (85%) were vacationers or occasional visitors who had little to no familiarity with rip currents; 90% had poor swimming skills. Mocellin recommended that lifeguards with appropriate lifesaving equipment be stationed along the central-west beaches in Santa Catarina. One result of beach safety programs in northern Santa Catarina was an 80% reduction of fatal accidents along 100 km of beaches (Figure 13.1).

In Rio Grande do Sul, the southernmost state of Brazil, Maia (2008) conducted a study of Cassino—a microtidal, dissipative, multi-bar beach by analyzing morpho-dynamic, video, and rescue data (Calliari and Klein, 1993). This approach allowed the identification of a wide variety of bar morphologies characterized by steep bars

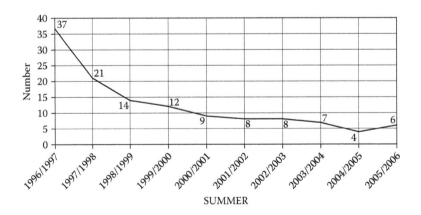

FIGURE 13.1   Number of fatal bathing accidents between 1996 and 2006 on beaches of Santa Catarina state. (*Sources:* Klein, A.H.F., Santana, G.G., Diehl, F.L. et al. 2003. *J. Coastal Res.*, 35: 107–116; Mocellin, O. 2006. Universidade do vale do Itajaí. Dissertação de Mestrado.)

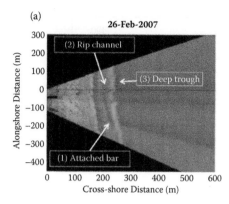
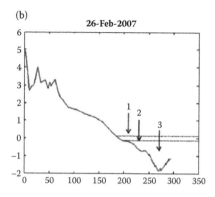

FIGURE 13.2 Image of Argus System (a) and beach profile (b). Horizontal line in (a) indicates a risky situation for bathers due to rip currents associated with surf-zone morphology. Bar attached to the beach (1), rip channel (2), and deep trough (3) are shown. (*Source:* Modified from Maia, N.Z. 2008. Universidade Federal do Rio Grande. Monografia de Conclusão de Curso. With permission.)

and deep troughs that can form, migrate, and be destroyed over short time intervals. The video images allowed the identification of morphological features generally associated with intermediate beach stages that may develop rip currents. Figure 13.2a shows a risky situation—an area with increased numbers of bathers during a summer season. The Timex image (average of 600 images) shows the first bar welded to the beach, followed by a deep trough, and then a second bar near an abrupt trough 2 m deep. During field work to obtain beach profiles, Maia (2008) recorded strong rip currents (Figure 13.2b). Rescue data obtained in the summers of 2006 and 2007 indicated that the higher number of accidents (77%) occurred preferentially between 2 and 7 pm, and the victims were primarily males aged 6 to 25 years old.

NORTHEAST BRAZIL

Futuro Beach in Fortaleza, the capital city of the state of Ceará, was studied by Albuquerque (2008) and Albuquerque et al. (2010). This 8-km beach is well known for high numbers of drownings when compared with other mesotidal beaches in the metropolitan area (Figure 13.3). The reasons for these losses are high numbers of bathers and frequent occurrences of moderate to strong rip currents. Field data consisted of in situ observation of morphodynamic characteristics and lifeguard data concerning beach bathing accidents.

Studies of beachgoers showed that the highest rates of accidents among males aged 21 to 28 years (59%). Most of the rescues (66%) occurred at an intermediate beach stage near low tide and were associated with transverse bars. Data from the Futuro Beach lifeguards showed that rip currents were the most serious hazards for bathers (86% of cases). During a Santa Catarina project with a similar approach by Klein et al. (2000), eight lifeguard stations were installed along Futuro Beach. The lifeguards were instructed to act preventively by advising the bathers when they approached risky areas. This action resulted in a reduction of the number of accidents

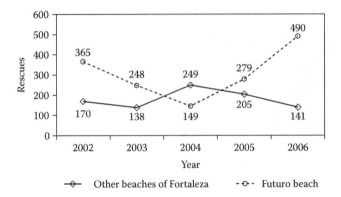

FIGURE 13.3   Number of rescues in Futuro beach when compared with other beaches in the metropolitan area of Fortaleza. (*Source:* Albuquerque, M.G. 2008. Universidade Federal do Rio Grande. Dissertação de Mestrado.)

by more than 50% (300 to 450 per year to 149) annually (Figure 13.3). After the deactivation of the lifeguard stations, the number of accidents increased again.

Albuquerque et al. (2010) found that drownings were more frequent where beach sediments were composed of medium to coarse sand. This was also noted by Calliari et al. (2010) along the northern beaches of Rio Grande do Sul state (Figure 13.4) who postulated that medium and coarse sand beaches exhibit greater mobilities and larger morphological changes. This combination apparently favors rip currents that are linked mainly to rhythmic and transverse bars (Wright and Short, 1984). It is important to note that even fine-grained, dissipative beaches characterized by low mobility can display morphological states that may be harmful to unskilled bathers.

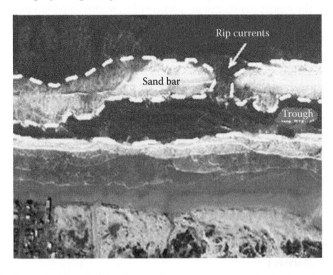

FIGURE 13.4   Aerial photos (ADAR 1000 System-LOG-FURG) displaying bar and rip currents at Capão da Canoa Beach, a medium sand, intermediate beach on the northern coast of Rio Grande do Sul state.

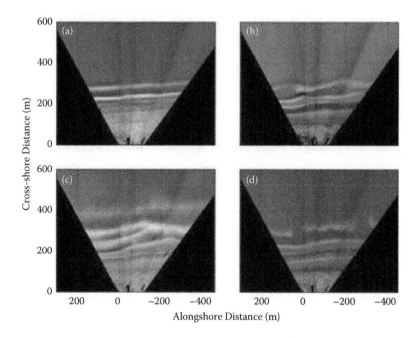

FIGURE 13.5 Long-term exposure of vídeo images generated by Argus system at Cassino Beach in 2005 showing longitudinal bar and trough (a), rhythmic bar and beach (b), irregular bars (c), and double bar system with potential to generate rip currents (d).

Figure 13.5 shows a series of video images obtained at Cassino Beach, illustrating different surf zone morphologies that have the potential to be rip-prone.

DISCUSSION AND CONCLUSIONS

The decrease in fatal accidents in Santa Catarina state (Hoefel and Klein,1998; Klein et al., 2003; Mocellin, 2006) and the lower numbers of rescues at the northeastern beaches (Albuquerque, 2008) after intense educational campaigns show that beach safety programs should take into account beach stage and public usage. Coastal development leads to increased beach use and therefore more public risk. Short and Hogan (1994) showed that changes in surf zone morphodynamics are predictable, and bathers should be classified according to age and swimming skills in order to devise a risk index (hazards + beach user characteristics). Such information can be used to mitigate risks through educational programs about beach hazards and by providing rescue equipment to lifeguards. Beach safety programs should be proactive and have appropriate resource allocations instead of merely being reactive to trauma and tragedy. Although programs utilizing this approach have been developed in some Brazilian states, many were discontinued.

Historical analysis of the rescue service and beach safety programs demonstrate that they can be successfully undertaken through the joint efforts of Firefighters Brigade staff or trained civilian personnel and institutions specializing in research on beach morphodynamics. In addition, Maia (2008) suggested that education about the risks inherent on surf beaches should start early. One recommendation is that surf beaches be monitored continuously by video systems such as the Argus (Holman et al., 1993; Holman and Stanley, 2007). Descriptors of beach hazards, such as bar morphology, breaker height, and presence of rip currents, can be identified in real time. Such data can be used to educate the public and diminish the occurrences of accidents and drownings. Coastal monitoring by video provides an easy and inexpensive way to collect data on beach conditions and numbers of users. This monitoring and management program is possible through collaboration between public authorities and educational institutions to help ensure safe beach use.

REFERENCES

Albuquerque, M.G. 2008. Morfodinâmica da praia do Futuro, Fortaleza–Ceará. Programa de Pós-Graduação em Oceanografia Física, Química e Geológica. Universidade Federal do Rio Grande. Dissertação de Mestrado.

Albuquerque, M.G., Calliari, L.J., and Pinheiro, L.S. 2010. Analysis of major risks associated with sea bathing at Futuro Beach, Fortaleza–Ceará, Brazil. *Braz. J. Aquatic Sci. Technol.*, 14: 1–8.

Angelotti, R. 2004. Segurança dos usuários de praia e riscos associados ao banho de mar em Pontal do Paraná. Universidade Federal do Paraná. Monografia de Conclusão de Curso.

Calliari, L.J. and Klein, A.H.F. 1993. Características morfodinâmicas e sedimentológicas das praias oceânicas entre Rio Grande e Chuí. *RS Pesquisas*, 20: 48–56.

Calliari, L.J., Guedes, R.M.C., Lélis, R.F. et al. 2010. Hazards and risks associated to coastal processes along the southern Brazilian coastline: a synthesis. *Braz. J. Aquatic Sci. Technol.*, 14: 51–63.

Goya, S.C. 2009. Mudanças climáticas e desafios ambientais. *Sci. Am. Brazil*, 2: 56–61.

Hoefel, F.G. and Klein, A.H.F. 1998. Environmental and social decision factors of beach safety in the central northern coast of Santa Catarina, Brazil. *Notas Tec. FACIMAR*, 2: 155–166.

Klein, A.H.F., Mocellin, O., Menezes, J.T. et al. 2005. Beach safety management on the coast of Santa Catarina, Brazil. *Zeitschr. Geomorphol.*, 141: 47–58.

Holman, R.A., Sallenger, A.H. Jr., Lippmann, T.C. et al. 1993. The application of video image processing to the study of nearshore processes. *Oceanography*, 6: 78–85.

Holman, R.A. and Stanley, J. 2007. The history, capabilities and future of Argus. *Coastal Eng.*, 54: 477–491.

Klein, A.H.F., Santana, G.G., Diehl, F.L. et al. 2003. Analysis of hazards associated with sea bathing: results of five years' work in oceanic beaches of Santa Catarina state, southern Brazil. *J. Coastal Res.*, 35: 107–116.

Klein, A.H.F., Santana, G.G., Diehl, F.L. et al. 2000. Análise dos riscos associados ao banho de mar: exemplos de praias catarinenses. In: Simpósio Brasileiro sobre Praias Arenosas: morfodinâmica, ecologia, usos, riscos e gestão. *UNIVALI*, 45–49.

Maia, N.Z. 2008. Riscos potenciais costeiros associados à segurança de banho na praia do Cassino, RS: análise de fatores morfodinâmicos e sociais. Universidade Federal do Rio Grande. Monografia de Conclusão de Curso.

Mocellin, O. 2006. Determinação do nível de risco público ao banho de mar das praias arenosas do litoral centro norte de Santa Catarina. Universidade do vale do Itajaí. Dissertação de Mestrado.

Pereira, P.S. and Calliari, L.J. 2005. Daily beach changes during the summers of 2002–2003 in the tourist terminal sector, Cassino beach, RS, Brazil. *Braz. J. Aquatic Sci. Technol.*, 9: 7–11.

Povh, D. 2000. Economic instruments for sustainable development in the Mediterranean region: responsible coastal zone management. *Period. Biologor.*, 102: 407–412.

Short, A.D. and Hogan, C.L. 1993. The Australian beach safety and management program: Surf Life Saving Australia's approach to beach safety and coastal planning. *11th Australasian Conference on Coastal and Ocean Engineering*, National Conference Publication 93/4: 113–118.

Short, A.D. and Hogan, C.L. 1994. Rip currents and beach hazards: their impact on public safety and implications for coastal management. *J. Coastal Res.*, 12: 197–209.

Souza, P.H. 2005. O serviço de guarda-vidas no litoral paranaense nas temporadas de 1997/1998 a 2004/2005. Especialização em Planejamento e Controle em Segurança Pública. Universidade Federal do Paraná. Monografia de Conclusão de Curso.

Szpilman, D. 2004. *Manual de Emergências Médicas*. Rio de Janeiro.

Wright, L.D. and Short, A.D. 1984. Morphodynamic variability of beaches and surf zones: a synthesis. *Marine Geol.*, 56: 92–118.

World Health Organization. 2009. Detailed data files of the WHO mortality database. http://www.who.int/whosis/mort/dowload/en/index.html

14 Rip Current Hazards on Large-Tidal Beaches in the United Kingdom

Timothy M. Scott, Paul Russell, Gerd Masselink, M. J. Austin, S. Wills, and A. Wooler

CONTENTS

INTRODUCTION

Beach hazards have historically been addressed in terms of damage to property and infrastructure. In recent years, scientists have started to develop an understanding of hazards and risks to beach users (Lushine, 1991; Short and Hogan, 1994; Leahy et al., 1996; Lascody, 1998; Short, 1999; Short, 2001; Engle et al., 2002; Hartmann, 2006; Sherker et al., 2008). Application of improved knowledge of beach morphodynamics and rip currents for beach safety was initiated in Australia, principally through the work of Short (1999). He applied beach state models to hazard assessment, leading to the development of inventories of beach types and hazards for all Australian beaches.

The Royal National Lifeboat Institution (RNLI), the principal provider of lifeguard services in the United Kingdom, commissioned research to further understanding of beach morphodynamics and hazards (Scott et al., 2007, 2008, 2009; Scott, 2009). The research outcomes provide a basis for the development of practical hazard assessment tools and improved lifeguard training and public education. These elements are integral parts of the RNLI risk assessment and mitigation program. One of the principal goals is to improve understanding of rip current hazards at medium- to high-energy macro-tidal beaches.

Rip currents have long been documented as significant hazards to waders and swimmers (Shepard, 1949; McKenzie, 1958; Short and Hogan, 1994; Short, 1999; and MacMahan et al. 2006). Lascody (1998) stated that rips in Florida on average caused more deaths than hurricanes, tropical storms, lightning, and tornadoes combined. Recent investigations of beach hazards in the UK, Australia and the United States indicate that rip currents represent the single most significant cause of rescues and fatalities for recreational beach users (Short and Brander, 1999; Scott et al., 2007; Scott et al., 2008). Specifically, Scott et al. (2008) noted that 68% of all incidents recorded by the RNLI on UK beaches were due to rips.

Most previous investigations of rip current hazards concerned micro- and meso-tidal environments (<4 m tidal range). The macro- to mega-tidal beaches (4 to 12 m tidal range) that dominate the UK coast introduce unique complexities into understanding beach hazards. In addition to large tides, the UK beach environment is characterized by a mixed, often high-energy, wave climate, and complex geological history.

The nature of rip currents and their spatial and temporal distribution and relationship with beach type and morphology are of prime concern (Scott, 2009). The aim of this chapter is to synthesize new insights regarding the controls on the temporal hazard signature (THS)—the variation in space and time in the types and severities of bathing hazards.

ENVIRONMENTAL SETTING

The UK by virtue of its location and geologic setting possesses a broad spectrum of beach environments along its 5,000-km shoreline. Its beaches attract large numbers of visitors annually due to their aesthetic, sport, and recreational appeal and they provide pivotal support to the tourism industry in many regions. Characterizing rips and recreational beach hazards in a UK setting requires a comprehensive understanding of physical beach environments.

The wide variety of beach systems throughout England and Wales is driven by the along-coast variability of static and dynamic environmental factors. The three most important factors are geology, sediments, and external forcing (wind, waves, storms, and tides). The spatial variability in boundary conditions is responsible for geographical variations in coastal morphology and morphodynamics (Davies, 1980) that in turn control the levels of physical hazards.

Steers (1960) attributed the diversity in coastal geomorphology in England and Wales mainly to varieties of rocks. The large-scale solid geology, characterized by a decrease in age and rock resistance from west to east, forms the template of the overall coastal topography and creates a contrast between the high-relief, mainly rocky, west coasts of England and Wales, and the low-relief, mainly unconsolidated, east coast of England (Clayton and Shamoon, 1998). Coastal sediments were largely derived from the most recent glaciations that deposited large quantities of hetero-geneous materials. The abundance of these coastal sediments significantly affects beach morphology, often constraining the extent of morphological evolution (Jackson et al., 2005) and affecting the hydrodynamic regime.

Beach sediments are transported mainly by tide- and wave-driven currents that exhibit large spatial variabilities (Figure 14.1). Most of the coasts experience macro- (41.7%) or mega-tidal tide ranges (42.2%), and the mean spring range (MSR) is 5.77 m. The largest tides (MSR >12 m) occur in the Bristol Channel due to the funneling effects of the coastal topography. The smallest tides (MSR = 1.2 m) are experienced in the lee of the Isle of Wight.

Some of the most energetic wave conditions are experienced southwest of the UK, where the mean significant wave height (H_s) is between 1.25 and 2.25 m and the wave climate is a mixture of Atlantic swell and locally generated wind waves (Figure 14.1b). The lowest wave conditions prevail in the northwest and east of England, where wind waves are predominant and mean H_s values are less than 1 and 1.25 m, respectively. Exposure of southwest England to the Atlantic Ocean increases the contributions of long-period swell waves to the wave spectrum. The complexities of coastal orientation and exposure around the coasts of England and Wales lead to beach waves that are a dynamic balance of high- and low-energy and wind-swell-wave components that are often characterized by a bi-modal wave energy spectrum with multiple directional sources (Bradbury et al., 2004).

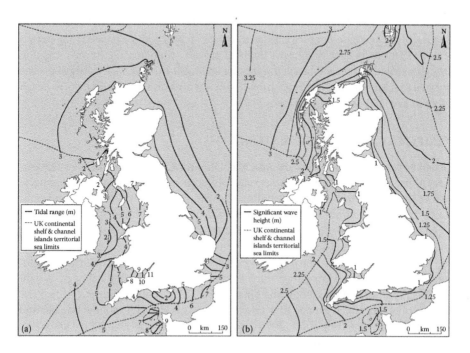

FIGURE 14.1 (a) Mean spring tide range (based on data derived from an average tidal year). (b) Annual mean significant wave height based on hourly model hindcast data over 7 years. (*Source:* Adapted from Department of Business Enterprise and Regulatory Reform, 2008).

FIGURE 14.2   (*See color insert.*) Crooklets Beach, north Cornwall provides an example of the complex nature of physical hazard dynamics. Low-tide rip systems controlled by transverse bar, and rip beach morphology and upper beach morphodynamics modified by geologic control and groundwater seepage, constrain and influence surf zone currents and hazards during mid and high tides. Inset shows location of southwest England study region (grey box) and Crooklets Beach location (solid circle). (Photo courtesy of Tim Scott.)

Mean seasonal variations in wave climates are significant in many coastal regions with strong summer to winter variations. Wave buoy data from the Atlantic southwest coast of England show that significant wave heights range from 2 to 5 m from summer to winter, respectively. Joint wave distributions indicate that a significant portion of the increase in energy is due to winter storms with associated long-period waves (T_m up to 14 sec). Southwest England provides a unique site for investigating the role of rip currents and beach hazards for sediment-limited, high-energy beaches with large tidal ranges. These beach environments present the greatest risks to beach users in England and Wales (Figure 14.2).

TEMPORAL HAZARD SIGNATURE (THS)

Beach hazards, like morphodynamics, vary on a range of time scales. For the successful provision of beach safety services, it is crucial to understand the temporal and spatial variabilities of the prevailing hazards. Figure 14.3 is a conceptual summary of the key findings within the context of a THS for high-risk bathing beaches. The framework provides a structure for beach hazard assessment as defined by beach type, environmental setting, and hydrodynamic forcing.

This approach stems from Short and Hogan (1994), where modal and wave height-modified beach hazard ratings were defined for each beach state described by Wright and Short (1984) and Masselink and Short (1993). This research initiative considers the intermediate beach groups [low tide terrace + rip (LTT+R) and low tide bar and rip (LTBR)] redefined by Scott (2009) as the most hazardous. Dissipative and ultra-dissipative beaches have relatively low hazard ratings because the absence of sand bars means low rip activity.

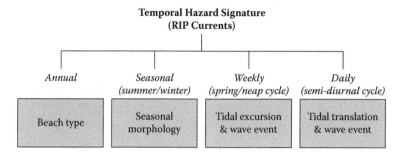

FIGURE 14.3 Time components of the temporal hazard signatures for the macro-tidal beaches studied. Principal environmental controls on rip current hazards associated with each time scale are indicated.

BEACH TYPE

The physical characteristics of 98 beaches in England and Wales were sampled, including morphology, sedimentology and hydrodynamics (Figure 14.4). Cluster analysis of this beach database produced a hazard classification comprised of twelve beach groups—each with a distinct hazard signature (Scott, 2009). This model provides the framework for assessment of the distribution of rip current morphology and characteristics of physical hazards at all 76 RNLI beaches that were actively patrolled from 2005 to 2007 (Figure 14.4).

Few researchers other than Short (1993, 1999 and 2001) and Short and Hogan (1994) included the concepts of morphological state and beach type in hazard evaluation. In many cases, only wave height, period, and direction were used (Lushine, 1991). Ten of the twelve identified beach groups are represented within RNLI lifeguard patrolled locations (Scott, 2009). Figure 14.5 shows the ten groups with idealized morphological forms and numbers of representative RNLI beaches for each group. These contrasting study environments provide a unique opportunity to identify specific hazard characteristics (Scott, 2009).

RNLI incident records showed that rip currents were the causal hazards for 68% of all reported incidents between 2005 and 2007. In particular, the high-energy, intermediate beaches with low-tide bars and rip morphology (LTT+R and LTBR) present the greatest rip current risk to recreational beach users. Eighty percent of all reported incidents on these beaches were due to rip currents (Figure 14.5). These high-risk beaches, representing 59% of the west coast beaches in Devon and Cornwall and 77% of all RNLI patrolled beaches, also attracted the greatest number of visitors (Figure 14.6).

The importance of spatial variations of wave energy levels is related to minimum wave energy thresholds for transport (Masselink and Short, 1993) and for the beach groupings defined herein as the threshold energy levels required for generating infragravity waves (Guza and Thornton, 1985) and rhythmic bar morphology. A critical wave energy threshold is defined as H_s of 0.8 m and T_p of 8 sec to separate

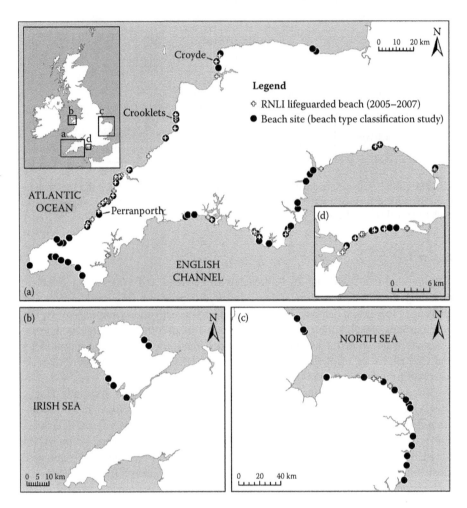

FIGURE 14.4   United Kingdom map indicating locations of all 98 beach sites included in the beach type classification study (black circles) and 76 beaches lifeguarded by the Royal National Lifeboat Institution from 2005 to 2007 (white crosses). Subplots and inset represent enlarged views of region indicated on the overview map.

low- and high-energy intermediate beach groups. This delineation separates beaches dominated by long-period, open-ocean swells from beaches with fetch-limited wind waves. Controlling for the presence or absence of bar and rip morphology, this distinction is key in understanding how hazard levels vary by beach type.

Jackson et al. (2005) and McNinch (2004) suggested that hard-rock geology acts to constrain and modify morphodynamic processes by controlling sediment abundance and depth to geologic substrate. The combination of beach type and environmental controls (geologic and structural constraint, sediment abundance, drainage, and backshore geomorphology) define the potential for rip current activity. Rips can take on a number of forms based on their forcing and controlling mechanisms

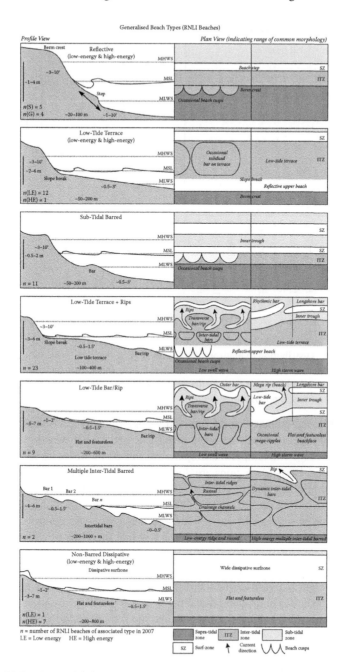

FIGURE 14.5 Generalized beach types (Scott, 2009) associated with Royal National Lifeboat Institution patrolled beaches in 2007. The number of beaches associated with each type (*n*) is indicated as are typical morphological features and commonly observed variabilities (plan view illustrations in right-hand panels).

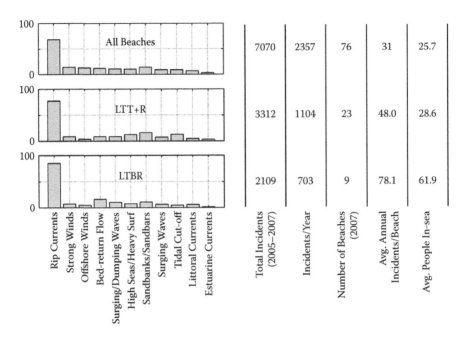

FIGURE 14.6 Distribution of environmental causes of hazards for Royal National Lifeboat Institution incidents at all beaches (top) and only LTT+R and LTBR beaches with rip morphologies (middle and bottom). Data collected from 2005 to 2007. Additional statistics provide incident and beach population numbers.

(Short, 1985). Accretionary beach rips and topographically controlled rips are the largest contributors to beach hazards for LTT+R and LTBR beaches (Scott et al., 2009; Figure 14.7). The temporal and spatial variations of these rips are due to large tidal excursions. Low-tide regions are dominated by beach rips within the sub- and low-tide rhythmic bar systems (Figure 14.7, left). During high-tide, these systems are often in >8 m water depth, and surf zone processes interact with a steeper upper beach (LTT+R examples). In cases with significant geologic control, topographic rips are located up to 500 m landward of the low-tide shoreline (see Figure 14.2 and Figure 14.7, right panel).

Unlike some of the more reflective or ultra-dissipative beach groups where large tidal range or sediment size restricts temporal state change, intermediate dynamic LTT+R and LTBR beaches exist around critical thresholds of the dimensionless fall velocity $\Omega = H_b/w_sT$, where H_b is breaking wave height, w_s is sediment fall velocity (related to sediment size), and T is wave period (Gourlay, 1968; Dean, 1973; Davidson and Turner, 2009). Variation of Ω around these threshold values is controlled by intra-annual wave conditions (wave steepness) and hence seasonal beach erosion and accretion.

SEASONAL (SUMMER AND WINTER) MORPHOLOGIES

Seasonal monitoring of hydrodynamics and morphology at LTT+R and LTBR beaches identified key mechanisms controlling the temporal hazard signature (THS).

Low-tide accretionary beach rips	Mid-tide accretionary beach rips	Mid/high-tide topographic rips
(Rhythmic bar morphology)	(Inter-tidal bars)	(Inter-tidal geology)

FIGURE 14.7 High-risk rip current types in the United Kingdom identified by this study as responsible for most rip-related incidents and rescues reported by the Royal National Lifeboat Institution.

Topographic surveys and video imagery documented intra-annual and seasonal morphological changes at LTT+R and LTBR beaches (annual H_s = 1.25 to 2.25 m; MSR = 4.2 to 8.6 m). Offshore sediment transport (below MLWS) and intertidal beach lowering occur during high-energy winter periods. These erosive periods create flat, featureless, intertidal zones and quasi-linear longshore bar and trough (LBT) sub-tidal bar systems. Figure 14.8 illustrates the subsequent accretion and re-establishment of rhythmic, and then transverse, lower inter-tidal bar and rip systems measured during the lower energy spring and summer period at Perranporth. The transition of beach morphology to an increasingly three-dimensional state under decreasing wave energy conditions as well established (Wright and Short, 1984).

Figure 14.8a shows time-averaged oblique (unprocessed) video imagery, indicating sub-tidal bar crest locations. Figure 14.8b illustrates the accretionary transitions through monthly collected topographic RTK-GPS survey data in conjunction with the ortho-rectified video images from Figure 14.8a. This sequence clearly shows the extent of bar development in the mid-tide region, the increasing three-dimensional nature of the low- and sub-tidal bar, and rip configurations leading to exposure of the accreting low-tide bar during the September 12, 2007 survey.

The cross-shore distribution (through tidal elevation) of incidents, normalized by frequency of tidal elevation, is shown in Figure 14.8c. During the early season (May to June), incidents are largely restricted to the low-tide region. The increased contributions of mid- and high-tide incidents to cross-shore distribution throughout the rest of the season (July to October) can be linked to the development of inter-tidal bars at Perranporth. The increasing bar morphology within the higher tidal regions

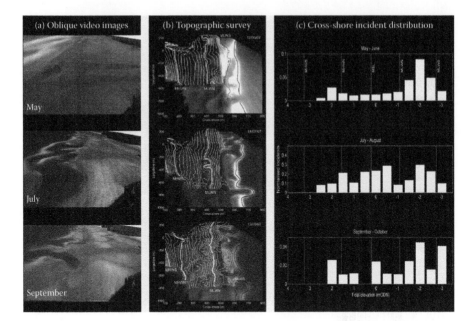

FIGURE 14.8 (a) Oblique time average video imagery and (b) topographic RTK GPS survey data combined with rectified time-averaged video imagery illustrate seasonal variations in bar morphology at Perranporth Beach in 2007. Perranporth represents a typical LTBR type; the mid and low tide bar transitions are similar to those seen at LTT+R beaches. Mean tidal elevations (bold white) are labeled and approximate location of the shoreline (black dashed) and the wave break point bar (black solid) are indicated. (c) Incident occurrence at Perranporth associated with vertical beach elevation provides an insight into the cross-shore locations of incidents. Incident occurrence at 0.5-m intervals is normalized by frequency of tidal elevation for early, mid, and late season. Dashed lines indicate mean tidal levels.

and progressive three-dimensional transition and onshore migration of the low- and sub-tidal bar systems during the summer accretionary waves extends rip activity through more of the tidal cycle and enhances the hazards. Magnification of temporal rip hazards in combination with greater beach populations during warmer summer waters resulted in an incident peak at Perranporth (Figure 14.8c).

Waves

Wave forcing over a dynamic bar morphology drives rips (Sonu, 1972; Haller et al., 2002) and is the principal component of rip prediction (Lushine, 1991; Engle et al., 2002). Thus, if waves drive rip circulation, how do event-scale variations in the wave climate (days to weeks) affect hazards, and do rip hazards increase linearly with wave height?

The joint wave distribution of the entire 2007 patrol season from nearshore wave buoy records (10 m depth) at Perranporth, west Cornwall show that the highest frequency wave heights were from 0.5 to 1 m with peak wave periods of 4 to 12 sec. Joint distribution clusters identify short-period, medium-energy events

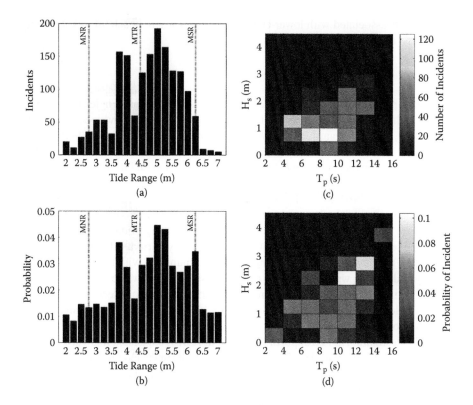

FIGURE 14.9 (*See color insert.*) Histograms of tidal range associated with (a) incident frequency and (b) probability of incident (*IR*). Dashed lines indicate mean spring range (MSR); mean tidal range (MTR) and mean neap range (MNR) are marked. Data represent incidents recorded at all studied west coast LTT+R and LTBR beaches in 2007. Two-dimensional frequency matrices of joint wave distribution are associated with (c) number of incidents and (d) probability of incident. (*Source:* Data recorded during patrol hours, 2007 patrol season.)

and long-period, low-energy events as most common. Incident counts show medium- to long-period, low-energy wave conditions (H_s of 0.5 to 1 m; T_p of 6 to 10 sec) associated with the highest number of rip incidents (241) at the six west Cornwall LTT+R and LTBR beaches (Figure 14.9). This represents 28% of all rip incidents at the selected beaches during the season. Unsurprisingly, the in-sea populations were largest during low-energy conditions ($H_s < 1.5$ m).

The probability of incident (*IR*) for recreational beach users is expressed as $IR = Re/P$ where P is the number of people in the water and Re is the number of individuals assisted or rescued. *IR* is highest when associated with high-energy wave conditions ($H_s > 2$ m) with peak periods >10 sec (Figure 14.9d). This reflects a small number of incidents that occur during the early season when hazardous high-energy conditions combine with low in-sea populations, leading to low levels of hazard exposure. About 75% of beach users were in the sea during low-energy wave conditions (H_s of 0.5 to 1.5 m); longer-period swells accounted for a large proportion (36%) of incidents under low-energy conditions. This highlights the importance of understanding hazard levels associated with lower-energy conditions where beach user exposure to rip hazards are high.

TIDES

Rip hazard prediction has begun to incorporate tidal modulation (Engle et al., 2002). Tidal level has been widely observed to modify rips, increasing flow speeds at lower tides (Sonu, 1972; Aagaard et al., 1997; Brander, 1999; Brander and Short, 2001; MacMahan et al., 2005). For macro-tidal beaches, tidal modulation of rip hazards is exerted through two principal mechanisms: tidal excursion and translation.

Tidal Excursion (Spring/Neap Cycle)

The spring/neap tidal cycle creates significant temporal variations in tide ranges at macro-tidal beaches that have wide intertidal zones and large tidal excursions. This is exemplified at Perranporth Beach where over a 7-day period, tide range can vary from 2 to 7 m between neap and spring tides. In conjunction with the template morphology and forcing wave conditions, tidal excursion controls wave breaking and rip activity. This significantly affects rip numbers and associated incidence, particularly within the low-tide bar and rip region (Figure 14.9a and b). At LTT+R and LTBR beaches, more incidents occur when the tide range is greater than the mean.

Field investigation at Perranporth (Austin et al., 2009) suggested that variations in rip activity can be expressed as down-state (neaps to springs) and up-state (springs to neaps) morphological transitions; this finding is similar to that observed by Brander (1999) using beach state descriptions of Wright and Short (1984), but as a function of tidal level controlling wave dissipation (Figure 14.10). At the MLWN level, the typical wave dissipation pattern represents a LBT/RBB system, and weak rip circulation with alongshore flows dominating. As tidal elevation decreases toward MLWS level, the wave dissipation pattern represents TBR morphology, and the combination of spatially variable wave dissipation and morphological constriction results in the strongest rip flows. During the lowest observed tides, the rip system became isolated, representing the TBR/LTT configuration. Within this model are two hazardous transitions: (1) falling tide from LBT/RBB to TBR (down-state) and (2) rising tide from LTT to TBR (up-state). The extent of tidal excursion during spring/neap cycles controls the extent of these transitions and hence the levels of low- and mid-tide rip activation.

Analysis of video images at Perranporth shows that changes in tidal range affect rip hazards on a daily basis. During a neap to spring tide transition, variations in tidal excursion from one day to the next were found to be sufficient for a rip current system to become active. This is particularly relevant if low-tide bar and rip morphology was well developed during the neap phase, but only becomes active during the subsequent spring phase. The assumption that rip current activity is closely associated with bar and rip morphology was tested at Perranporth using GPS drifters (Austin et al., 2009; Austin et al., 2010; Figure 14.11).

Rip circulation and flow speeds shown in Figure 14.11 were recorded when low energy swell waves, spring tides, and well-developed bar and rip morphology

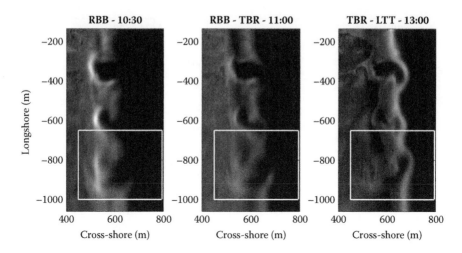

FIGURE 14.10 Time-averaged plan view video images of Perranporth Beach, west Cornwall, taken August 3, 2008 at the times indicated above panels (cross-shore distance increases to seaward). High pixel intensities (whiter regions) indicate high wave dissipation (breaking). Panels show progressive change in effective morphological beach state from RBB to TBR/LTT during falling tide. The highlighted boxes indicate the regions of GPS drifter deployment shown in Figure 14.11.

dominated. This indicates that rips can indeed switch on and off according to tidal elevation. Figure 14.11 shows all measured GPS drifter data classified into A_m/A_r bins representing degrees of morphological constraint on the rip flow. A_r is the cross-sectional area of the rip channel available for rip flow (increasing with tidal elevation), and A_m represents the morphologically constrained area of the channel (calculated at a water level where adjacent bars become exposed). A_m/A_r increases with decreasing tidal elevation as rip flow becomes increasingly channelized (Austin et al., 2010).

Currents recorded by the drifters were largely alongshore-dominated during higher tidal elevations associated with the RBB state in Figure 14.10. During the TBR and LLT stages, strongly rotational eddies constituted the principal circulation. Similar rotational circulation patterns were observed by MacMahan et al. (2010) within micro- and meso-tidal environments. This finding contradicts the long-standing notion of rip currents as seaward-flowing jets that expel bathers from the surf zone. In addition, alongshore-directed rip flows over the bar edge were of equal or greater speed than the offshore-flowing rip neck—an additional hazard. Under these conditions bathers standing on "safe" bar crests can be pulled laterally into a rip channel.

Tidal Translation (Semi-Diurnal Cycle)

Tidal translation is the rate of change in shoreline location and is key to developing high-risk conditions (Scott et al., 2009). While tidal excursion defines the extent of surf zone migration throughout the tidal cycle, tidal translation controls the rate of change of the tidal level and associated rip hazards. Many UK beaches have a

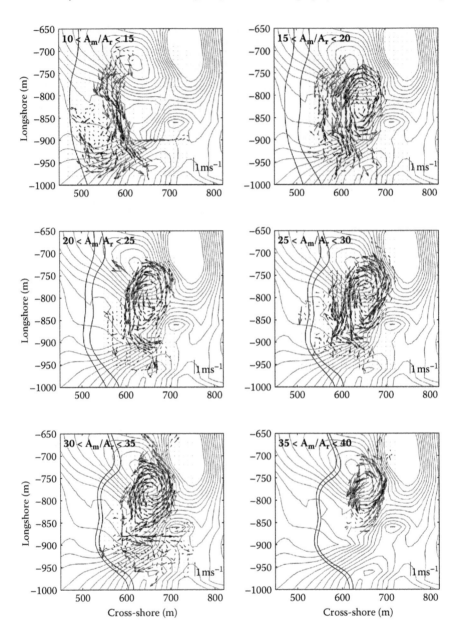

FIGURE 14.11 Mean Lagrangian GPS drifter circulation separated into classes defined by A_m/A_r that represent varying degrees of morphological constriction controlled by tidal level. Black vector arrows indicate rip speeds for bins classified as statistically significant (more than five independent observations). Gray arrows indicate all observations. Contours represent residual morphology. Black contours show approximate shoreline positional range for each bin. (*Source:* Austin, M., Scott, T., Brown, J. et al. 2010. *Cont. Shelf Res.*, 30: 1149–1165. With permission.)

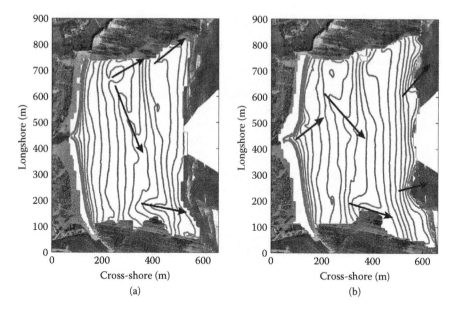

FIGURE 14.12   Fifteen-minute shoreline data (gray contours) from topographic surveys collected on spring tides on (a) June 18, 2007 and (b) September 14, 2007 at Croyde Bay, North Devon (*Source:* Scott, T.M., Russell, P.E., Masselink, G. et al. 2008. *Proc. Intl. Conf. on Coastal Engineering,* Hamburg, pp. 4250–4262.) Black arrows indicate regions of heightened rip current hazards during tides (*Sources:* Royal National Lifeboat Institution, Croyde Bay lifeguards, personal communication; background aerial images courtesy of Channel Coastal Observatory.)

tidal range >6 m with horizontal tidal excursions up to 600 m. For example, translation rates at Croyde Bay during large spring tides can reach 100 m in 15 min (Figure 14.12); this has significant implications for beach safety because of rapid alongshore migration of rips throughout the tidal cycle. In addition, large variations between spring and neap mean tidal excursion and translation can vary significantly on a daily basis.

HIGH-RISK SCENARIOS

Environmental factors as discussed above control the THS and combine to create high-risk scenarios that lead to mass rescue events as identified from lifeguard incident reports (Scott et al., 2009). During the 2007 patrol season, six mass rescue events occurred, significantly exceeding the coast-wide seasonal trend in *IR*. These events all involved more than 62 rip incidents in a single day (maximum of 151 per day) and more than 10 beaches simultaneously (maximum of 15). Examination of wave and tidal conditions as well as morphological state and beach population levels provides insight into key environmental conditions that cause high-risk situations for beach users.

Key Environmental Conditions

A number of key environmental conditions typically increase risk levels during periods of high-risk exposure (busy summer months):

- *Accretionary morphological conditions:* These conditions are observed on high-risk LTBR and LTT+R beaches with well-developed transverse bar systems; they evolve as a result of extended accretionary conditions during spring and summer months and are key elements in the creation of high rip risk levels for beachgoers.
- *Low- and medium-energy, long period swell waves:* Under these conditions often associated with summer accretionary periods, waves shoal on inner transverse bars during low- and mid-tides, generating strong alongshore variations in wave breaking and driving relatively strong rip flows. Rips that are morphologically constrained can have high flow speeds in relation to the forcing wave energy because of channelized flow between bars. These typical summer, low-energy conditions (H_s of 0.5 to 1 m; T_p of 6 to 12 sec) commonly occur in conjunction with high in-sea populations. These conditions allow greater bather activity in the surf zone, increasing exposure to rip current hazards.
- *Spring tides:* For macro-tidal regions, spring tidal periods are associated with increased rip incident risk through exposure of low-tide bars and rip current activation. Increased tidal excursion creates high tidal translation rates and hence rapid changes in rip hazards. The daily change of tidal range throughout the spring/neap cycle is also a significant control on the temporal hazard signature (THS) in macro-tidal regions where hazard levels change dramatically, even under similar wave conditions and beach morphology.

Practical Implications for Lifeguards

Mass rescue events are largely due to tidal changes that increase hazards for beach users. Tidal modulation of rip current circulation over well-developed bar morphology switches rips on and off through fluctuations in water levels. Tidal modulation of rips occurs at spring/neap (weekly) and semi-diurnal (daily) time scales. Complexity is added by intertidal transitions of rip locations both along- and cross-shore as tidal water levels change. These intertidal transitions can occur rapidly in macro-tidal environments and hazard exposure increases when the subsequent rip systems migrate alongshore, changing safe bathing areas into regions of rip activity. When this occurs, lifeguards must respond quickly by moving the designated bathing area laterally with the changing tide to regions of lower rip hazards.

Rip current circulation patterns are also modulated by tidal elevation. Strong alongshore flows can propel bathers from bar crests into rip channels. This is particularly problematic with large numbers of beach users. The circulatory nature of rips was observed, particularly during low tidal elevations, when only 10 to 20% of tracked drifters exited the surf zone. Instead of drifters being carried offshore into deep water, they circulated back over the bar—multiple times in some cases (Austin et al., 2010). Exposure of the low-tide bar crest during spring tidal conditions attracts bathers. During the subsequent flood, the bar crest rapidly submerges, and the feeder channel and rip system become active (TBR stage) as the in-sea beach user population passes through the feeder channel when trying to return to the beach.

Rapid changes in surf conditions due to high tidal translation rates require constant risk assessment and mitigation. Under certain combinations of events, the THS changes faster than reactionary mitigation measures can be implemented, resulting in coast-wide, mass rescue events.

The fundamental processes driving the complexities of rip currents (circulation patterns, flow speeds, and tidal modulation) are rarely well understood by even experienced lifeguards. With improved scientific understanding of surf zone processes and the temporal hazard signature, UK lifeguards will acquire the ability to provide more predictive and less reactive beach safety services.

CONCLUSIONS

This chapter discussed the nature of rip current hazards for large-tidal beaches in the UK. This work has shown that the temporal hazard signature, defined as the spatial and temporal variations in rip hazard characteristics, is controlled by a number of environmental factors. The combination of these factors creates high-risk scenarios that drive observed coast-wide mass rescue events. A basic scientific understanding of these mechanisms would equip lifeguards with additional knowledge required to improve risk mitigation during these high-risk scenarios.

KEY FACTORS CONTROLLING RIP HAZARDS

- *Beach type:* Characteristic beach morphology associated with different beach types is linked to the types and severities of expected rip current hazards. High rip hazards are associated with low tide terrace and rip (LTT+R) and low tide bar and rip (LTBR) beach types that commonly have low-tide bar and rip systems.

- *Seasonal beach change:* Intermediate LTT+R and LTBR beaches show significant seasonal changes in sand bar morphology from flat and featureless, erosive winter storm-dominated conditions to accretionary summer, swell-dominated conditions. During this transition, beach rip morphology commonly develops throughout the lower beach, increasing rip hazards with onset of the summer tourist season and warmer waters.

- *Waves:* Rip incident risk is highest during high-energy wave events that often occur in the early season when in-sea populations are low. High exposure of in-sea beach users (75%) during summer, low-energy, long-period wave conditions (H_s of 0.5 to 1.5 m) led to the highest number of rip incidents, illustrating the importance of understanding low- to medium-energy rip systems.

- *Tides:* Tidal modulation of water levels in large-tidal beaches exerts significant effects on exposure to rip hazards. Both the spring/neap and daily tidal cycles control the extent of tidal sweep (excursion) and the rate at which the shoreline position changes (translation). These two factors control whether a rip current system becomes active at low-tide or not and also control the rates of change of rip hazards as the surf zone moves from one rip system to another as the tide floods and ebbs. The complex combination of these effects controls hazard levels over seasonal, fortnightly, and daily time scales.

REFERENCES

Aagaard, T., Greenwood, B., and Nielsen, J. 1997. Mean currents and sediment transport in a rip channel. *Marine Geol.*, 140: 25–45.

Austin, M., Scott, T., Brown, J. et al. 2010. Temporal observations of rip current circulation on a macro-tidal beach. *Cont. Shelf Res.*, 30: 1149–1165.

Austin, M.J., Scott, T.M., Brown, J.W. et al. 2009. Macrotidal rip current experiment: circulation and dynamics. *J. Coastal Res.*, 56: 24–28.

Bradbury, A.P., Mason, T.E., and Holt, M.W. 2004. Comparison of the Met Office UK Waters Wave Model with a network of shallow water moored buoy data. *Proc. 8th Intl. Workshop on Wave Hindcasting and Forecasting*, Hawaii.

Brander, R.W. 1999. Field observations on the morphodynamics evolution of a low-energy rip current system. *Marine Geol.*, 157: 199–217.

Brander, R.W. and Short, A.D. 2001. Flow kinematics of low-energy rip current systems. *J. Coastal Res.*, 17:468–481.

Clayton, K. and Shamoon, N. 1998. New approach to the relief of Great Britain II. A classification of rocks based on relative resistance to denudation. *Geomorphology*, 25: 155–171.

Davidson, M.A. and Turner, I.L. 2009. A behavioural-template beach profile model for predicting seasonal to interannual shoreline evolution. *J. Geophys. Res.*, 114: F01020.

Davies, J.L. 1980. *Geographical Variation in Coastal Development*. Longman, New York.

Dean, R.G. 1973. Heuristic models of sand transport in the surf zone. *Proc. 1st Austral. Conf. on Coastal Engineering, Engineering Dynamics in the Surf Zone*, pp. 208–214.

Department for Business Enterprise and Regulatory Reform. 2008. *Atlas of Marine Renewable Resources*. Technical Report R1342. ABP Marine Environmental Research Ltd.

Engle, J., MacMahan, J., Thieke, R.J. et al. 2002. Formulation of a rip current predictive index using rescue data. In *Proc. 15th Annu. Natl. Conf. on Beach Preservation Technology*, Tait, L., Ed., Florida Shore & Beach Preservation Association, Biloxi, MS, pp. 285–298.

Gourlay, M.R. 1968. *Beach and Dune Erosion Tests*. Report m935/m936, Delft Hydraulics Laboratory, Delft.

Guza, R.T. and Thornton, E.B. 1985. Observations of surf beat. *J. Geophys. Res.*, 90: 3161–3171.

Haller, M., Dalrymple, R., and Svendsen, I.A. 2002. Experimental study of nearshore dynamics on a barred beach with rip channels. *J. Geophys. Res.*, 107: 1–21.

Hartmann, D. 2006. Drowning and beach safety management along the Mediterranean beaches of Israel: a long-term perspective. *J. Coastal Res.*, 22: 1505–1514.

Jackson, D.W.T., Cooper, J.A.G., and del Rio, L. 2005. Geological control of beach morphodynamic state. *Marine Geol.*, 216: 297–314.

Lascody, R.L. 1998. East Central Florida rip current program. *Natl. Weather Dig.*, 22 :2.

Leahy, S., McLeod, K., and Short, A.D. 1996. *Beach Management Plan*. Surf Life Saving Australia Ltd., Sydney.

Lushine, J.B. 1991. A study of rip current drownings and related weather factors. *Natl. Weather Dig.*, 16: 13–19.

MacMahan, J., Brown, J.W., Reniers, A. et al. 2010. Mean Lagrangian flow behavior on a open coast rip channeled beach. *Marine Geol.*, 268: 1–15.

MacMahan, J.H., Thornton, E.B., and Reniers, A. 2006. Rip current review. *Coastal Eng.*, 53: 191–208.

MacMahan, J.H., Thornton, E.B., Stanton, T.P. et al. 2005. RIPEX: observations of a rip current system. *Marine Geol.*, 218: 113–134.

McNinch, J.E. 2004. Geologic control in the nearshore: shore-oblique sandbars and shoreline erosional hot spots, Mid-Atlantic Bight, USA. *Marine Geol.*, 211: 121–141.

Masselink, G. and Short, A.D. 1993. The effect of tide range on beach morphodynamics and morphology: a conceptual beach model. *J. Coastal Res.*, 9: 785–800.

McKenzie, P. 1958. Rip current systems. *J. Geol.*, 66: 103–113.

Scott, T.M., Russell, P.E., Masselink, G. et al. 2009. Rip current variability and hazard along a macrotidal coast. *J. Coastal Res.*, 56: 895–899.

Scott, T.M., Russell, P.E., Masselink, G. et al. 2008. High volume sediment transport and its implications for recreational beach risk. *Proc. Intl. Conf. on Coastal Engineering*, Hamburg, pp. 4250–4262.

Scott, T.M., Russell, P.E., Masselink, G. et al. 2007. Beach rescue statistics and their relation to nearshore morphology and hazards: a case study for southwest England. *J. Coastal Res.*, 50: 1–6.

Scott, T.M. 2009. Beach morphodynamics and associated hazards in the UK. PhD dissertation, University of Plymouth.

Shepard, F.P. 1949. Dangerous currents in the surf. *Phys. Today*, 2: 20–29.

Sherker, S., Brander, R., Finch, C. et al. 2008. Why Australia needs an effective national campaign to reduce coastal drowning. *J. Sci. Med. Sport*, 11: 81–83.

Short, A.D. 2006. Australian beach systems: nature and distribution. *J. Coastal Res.*, 22: 11–27.

Short, A.D. 2001. *Beaches of the Southern Australian Coast and Kangaroo Island: Australian Beach Safety and Management Project.* Sydney University Press, Sydney.

Short, A.D., Ed. 1999. *Handbook of Beach and Shoreface Morphodynamics.* John Wiley & Sons, New York.

Short, A.D. and R.W. Brander. 1999. Regional variations in rip density. *J. Coastal Res.*, 15(3): 813–822.

Short, A.D. 1993. *Beaches of The New South Wales Coast: Australian Beach Safety and Management Project.* Sydney University Press, Sydney.

Short, A.D. and Hogan, C.L. 1994. Rip currents and beach hazards: their impact on public safety and implications for coastal management. *J. Coastal Res.*, 12: 197–209.

Short, A.D. 1985. Rip-current type, spacing and persistence, Narrabeen Beach, *Marine Geol.*, 65(1–2): 47–71.

Sonu, C.J. 1972. Field observation of nearshore circulation and meandering currents. *J. Geophys. Res.*, 77: 3232.

Steers, J. 1960. *The Coast of England and Wales in Pictures.* Cambridge University Press, London.

Wright, L.D. and Short, A.D. 1984. Morphodynamic variability of surf zones and beaches: a synthesis. *Marine Geol.* 56: 93–118.

15 Tracing Sand Movement in Strong Japanese Rip Currents

Nicholas C. Kraus

CONTENTS

INTRODUCTION

As part of a series of intensive field data collection studies in Japan (Horikawa and Hattori, 1987), several short- and long-term fluorescent tracer studies of longshore sand transport were conducted (Kraus, Farinato, and Horikawa, 1981; Kraus et al., 1982, Kraus, 1988). Prior to making these field measurements, it was standard procedure to reconnoiter the nearshore to avoid rip currents. In the early afternoon of August 21, 1980, a strong rip current was identified in the planned experiment area facing the Pacific Ocean at Ajigaura Beach, Japan, and it was decided to perform a trial sand tracer experiment.

In this pilot experiment conducted near mid tide, simultaneous measurements of the current and sand movement were attempted. Although this experiment was performed several decades ago, this previously unpublished research offers insights into the dynamics of rip currents and field measurement techniques, and the experiment may have been the first one attempting to measure sand transport synoptically in and around a rip current.

The reconnaissance surveys of the water circulation pattern were accomplished by hand-tethered floats (Sasaki and Horikawa, 1978) that indicated a persistent rip current throughout the day. The measurement procedure involved a tethered float or drogue (Figure 15.1). Upon a diver's signal, a float tied to the diver's wrist was

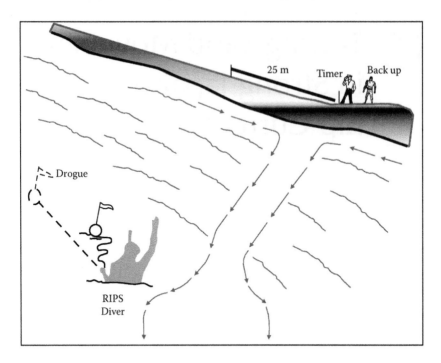

25 m   Timer   Back up

Drogue

RIPS
Diver

FIGURE 15.1   Reconnaissance survey of nearshore circulation by tethered float.

released and allowed to flow with the current. After the cord reached its 2-m extension, the diver signaled the timekeeper–recorder on shore, who also noted the direction according to the diver's hand signals. One or two stand-by divers accompanied the timekeeper as a safety precaution and back-up. The timekeeper moved along the shore at an interval marked by a chain or rope upon which a weight was attached to fix the initial position, pulling the weight to another position. The drogues were designed to sample the current about 0.3 to 0.6 m under the water surface by attaching fins (Figure 15.2) or by the shape of the drogue to minimize motion induced by wind and waves.

Professional salvage divers and the author, all in complete wet suits because of the cold water, placed instruments, injected tracers, and sampled the bottom while ropes were tied around their waists and fastened to screw anchors on shore (Figure 15.3). The ropes prevented divers from being swept to sea by rips that would have eliminated them from participating until they could swim around the rips and back to the site. A bathymetry survey was conducted in the same manner, with rod holders tied to shore by lines.

Eight electromagnetic current meters were deployed, of which five returned usable records. A capacitance wave gauge placed in the breakers malfunctioned, but an 8-mm memo-motion movie camera (triggered by an electric timer at fixed intervals) gave some quantitative indications of wave height. Three colors of sand tracers were injected and ten fixed stations were sampled at 15-min intervals for 180 min. The sand cores were split into segments and tracer grains counted under ultraviolet

FIGURE 15.2   Drogue with fins for sampling current below water surface.

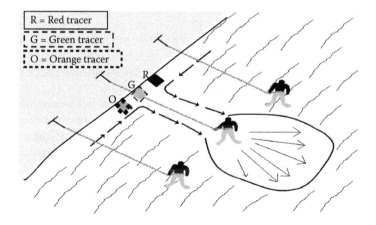

FIGURE 15.3   Divers tethered to shore for sediment sampling after tracer injection.

light. Depth of disturbance as calculated by Kraus (1985) and tracer movement could then be determined. The greatest depth of disturbance was found at the root of the rip, where the feeder currents turned to flow offshore. Apparent pulsations in tracer movement were observed at 45- to 60-minute intervals and interpretations made of the mode of transport as bed load or suspended load.

RIP CURRENT DYNAMICS: A SHORT REVIEW

A rip current is a strong and narrow seaward-directed flow of water in the surf zone. Rip currents are recognized as significant mechanisms for offshore sand transport of sediment and beach erosion (Komar, 1998) and are significant safety concerns for bathers. According to the heuristic model of Sasaki (1980), the dominant mode of offshore transport in rip currents is suspended load. However, no quantitative field study is known to have evaluated the mode and amount of sediment transported by a rip current or related rip current properties to sediment transport. This lack of information of rips obviously arises from the extreme difficulty and danger of operating in and around such powerful currents. Nevertheless, through experience gained in working in the surf zone as described here, it is possible to take a quantitative approach to study the hydrodynamics and sediment transport associated with rip currents.

The pattern of water flow associated with rip currents has been studied both theoretically and through field and laboratory observations. The basic phenomenon was described in the pioneering field studies of rip currents and nearshore circulation made by Shepard and coworkers (Shepard, 1936; Shepard et al., 1941; Shepard and Inman, 1950). Harris (1967) and Sonu (1972) classified rip currents within the context of observed nearshore flow patterns. Seminal sources for the numerical modeling of rip currents are the works of Arthur (1962) and Bowen (1969). One of the interesting features of rip currents is the long-period fluctuation in current velocity and direction. Dalrymple (1978) reviewed the generating mechanisms of rip currents. MacMahan, Thornton, and Reniers (2006) wrote a comprehensive review of rip current processes, and Yu and Slinn (2003) presented a modern numerical model of the wave–current interactions associated with rip currents.

On a long natural beach, rip currents may be considered transient perturbations of the surf zone longshore flow pattern and magnitude. Their influence can be neglected in determining the long-term longshore sediment transport rate. In contrast, near structures such as groins or jetties, the presence of rip currents is almost independent of the angle of wave approach. In such cases, the rip current is not a perturbation, but rather the major mechanism controlling sediment movement in the area. Sediment transported offshore by such a rip current is then available for further transport by any coastal current and may then be reintroduced to the littoral system on the other side of the structure, deposited in adjacent harbor basins, or lost offshore. Sasaki (1980) hypothesized that on beaches where normally incident waves and hence rip currents are dominant (Harris, 1967; Sonu, 1972), the short-term net longshore transport rate may be less than expected because of the interruption through offshore transport of sediment by rips. This sediment is returned to the surf zone during calmer weather.

Studies of sediment transport by rip currents and associated field data collection are thus necessary for improving both long- and short-term quantitative models of beach change and for making predictions about the genesis, persistence, location, strength, and dimensions of rips to promote beach bathing safety. Analysis of the complex environment of the rip current and coexistence of strong currents both parallel and perpendicular to wave incidence will yield valuable information for three-dimensional modeling of the nearshore circulation and resultant sediment transport. Finally, even after a rip current ceases to be active, relic features in the bathymetry, especially the rip channel, will continue to influence circulation patterns and sediment transport until the bottom is molded into a state compatible with the existing wave and flow conditions.

EXPERIMENT PROCEDURE AND LAYOUT

After a tethered float located a rip, the circulation pattern was determined by injecting a small amount of rhodamine dye near the shoreline. With the position of the rip current established, eight electromagnetic current meters were deployed in a symmetric array to record the rip feeder currents from both sides, current through the channel, and the current offshore. The instruments were ultimately set with a southward bias as shown in Figure 15.4 because of the strong current in that direction. Meter heads were positioned approximately 20 cm from the bottom. The current pattern was periodically observed with injected dye.

Although the plan was to install several capacitance type wave gauges at the site, rough wave conditions limited the number of wave gauges to one that unfortunately failed. A calibration check of the wave gauge was made at intervals by filming with a 16-mm memo-motion camera (Hotta, Mizuguchi, and Isobe, 1982). Wave height was determined from these films.

Sand transport was observed by injection of fluorescent tracer (dyed sand taken from the beach) and periodic sampling. Because of the acute difficulty for divers

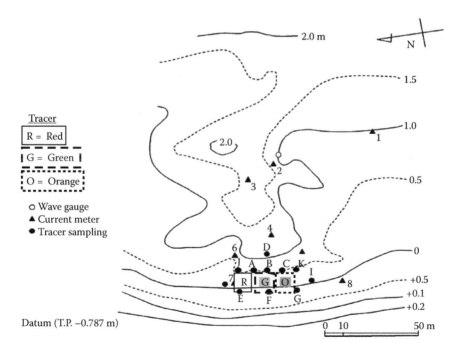

FIGURE 15.4 Instrument layout, tracer injection points, and tracer sampling stations around rip channel. The vertical datum is local mean sea level. T.P. (Tokyo Piel) = standard vertical water level datum in Japan (Tokyo Bay).

to move about in a rip current, the usual spatial sampling method was infeasible. A temporal sampling method (Kraus et al., 1982) requiring relatively few sampling points remained viable. Strictly speaking, tracer techniques are limited to situations where no erosion or accretion takes place. This is certainly not the case in a rip current. However, by taking large numbers of core samples, it is believed that temporal sampling may be capable of yielding an estimate of average transport, because, for example, a layer of fresh sand is distinguishable if observed over a layer of tracer.

Rapid sampling under such adverse conditions was made possible by a specially designed core sampler (Kraus, Farinato, and Horikawa, 1981). Unfortunately, because instrument setting took a significant amount of time, the suitability of the tracer method could not be fully tested because the sand transport portion of the experiment had to be shortened. Fading daylight halted tracer sampling after 2.5 hr of a scheduled 4-hr experiment. Temporal sampling designed around the one-time injection of tracer requires several hours. Duane and James (1980) presented evidence through example that a continuous injection method can reduce the sampling time to <1 hr. Such experiments seem suited to the rapidly changing conditions of the surf zone and rip currents, but are not explored further here.

Tracers of three colors (15 kg each) were injected as three point sources on a line parallel to the shoreline and approximately perpendicular to the axis of the rip channel (Figure 15.4). Three colors were placed in an attempt to correlate sand movements with local current strength. Stations J, A, B, C, and K formed the principal measurement line on which samples were taken every 15 min. Stations E, F, G, H, and I, sampled at 30-min intervals, served as checks for the general containment and movement of the tracers onshore and alongshore. Samples from Station D taken at 15-min intervals further indicated movement of tracers offshore.

Sampling stations could not be marked by poles because of the strong current and scour that would be induced around the poles. Divers located their sites by moving offshore to the full extension of a line of fixed length. These lines were anchored on the beach and also served as safety lines. All stations were sampled simultaneously. A tracer was not injected symmetrically with respect to the sampling grid because the rip current shifted slightly northward after the sampling stations were set, as observed from dye movement. The injection points thus shifted northward. A bottom topography survey proceeded simultaneously with the tracer experiment.

In the laboratory, core samples were cut into 2-cm segments up to 14 cm (Stations A, B, C, D, and F), or into 4-cm segments up to 12 cm (Stations E, G, H, I, J, and K). The segments were dried and weighed, and the number of tagged grains counted for each color. The result was then expressed as a concentration (tracer grains of a particular color/100g of sample).

RESULTS

CURRENT

Figure 15.5 depicts data from the operational current meters. The meters numbered 5 and 6 failed completely. The cross-shore u components of meters 1 and 7 and the

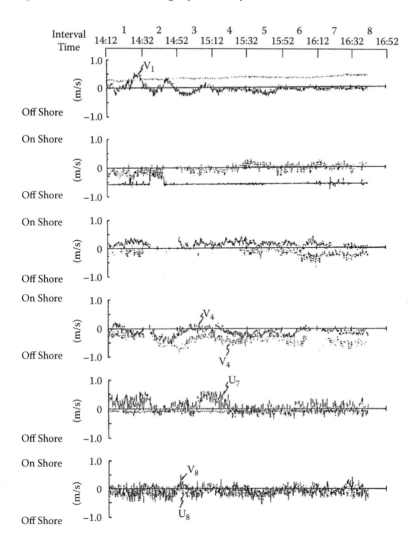

FIGURE 15.5 Raw current meter records. On-shore is positive; off-shore is negative.

longshore v component of meter 2 were also not available due to instrument malfunction. The current meter data indicate the existence of a long-period fluctuation in the cross-shore current velocity. To interpret the time change in the overall current pattern, averages were computed for eight intervals of approximately 20-min length. Figure 15.6 summarizes results for the current at approximately 20-min intervals.

TRACER CONCENTRATION

The tracer concentrations found at the sampling stations are displayed in Figures 15.7 through 15.12. The concentrations (y-axis) are shown on log scale. Plots on the right

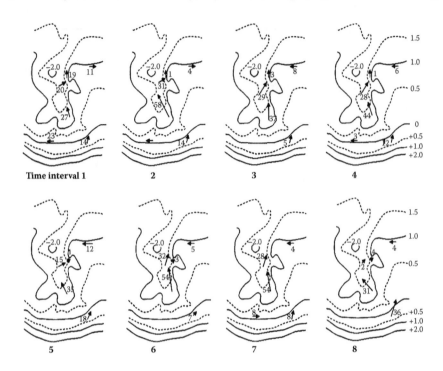

FIGURE 15.6 Available current averaged over approximately 20-min intervals for eight time segments. Values of current velocity are shown in centimeters per second; depth is shown in meters to approximately mean sea level.

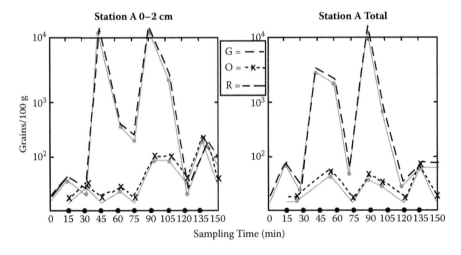

FIGURE 15.7 Tracer concentration through time slightly north and seaward of rip channel. No red tracer was observed at Station A.

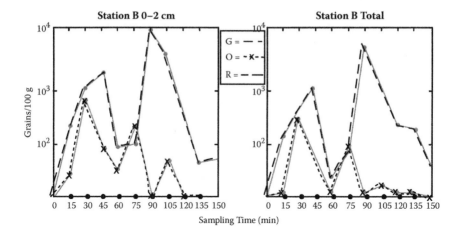

FIGURE 15.8 Tracer concentration through time seaward of rip channel. No red tracer was observed at Station B.

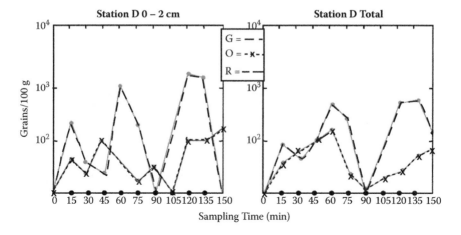

FIGURE 15.9 Tracer concentration through time offshore of injection area. No red tracer was observed at Station D.

show concentrations averaged over all segments in a core; those on the left indicate concentrations found in the top segment only (2- or 4-cm thickness). Comparison indicated that the tracer found in the top 2 cm was not always sufficient to represent the total concentration at a site. A sample consisting of the upper 4 cm more closely coincided with the total.

The most apparent feature in Figures 15.7 through 15.12 is the long-period fluctuation in concentration, comprising order of magnitude differences, found for both the green and orange tracers. Peaks are approximately 45 to 60 min apart. The first strong peak occurred earliest at Stations J and K. Station D showed early arrival of a small amount of tracer. The next and larger two pulses in tracer at

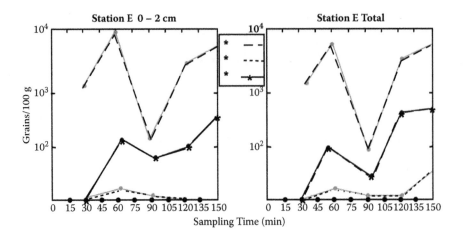

FIGURE 15.10   Tracer concentration through time north and shoreward of injection site.

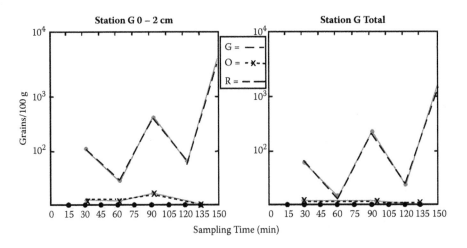

FIGURE 15.11   Tracer concentration through time south and shoreward of injection site. No red tracer was observed at Station G.

Station D arrived about 15 min later than the corresponding pulse at other stations; this is compatible with the differences in distances between the stations and injection point.

Because of the consistency in peaks among all stations and time lags in peaks at the most distant station from the injection line, we concluded that the pulses were actual and not random deviations. A similar pulsation in longshore transport was reported by Kraus, Farinato, and Horikawa (1981). These pulsations correspond to the long-period fluctuations in the waves and current passing over a bed. Note that the peak in concentration observed at Station B at 90 min was apparently transported to Station D in the throat of the rip, appearing 30 min later (120 min).

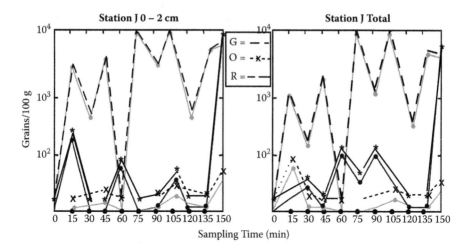

FIGURE 15.12 Tracer concentration through time, located north and seaward of injection site.

The centrally placed green colored tracer was injected directly at the base of the rip neck, as judged by the offshore movement of dye. Green tracer dominated and appeared in relatively large amounts at all sampling stations. The movement of the total sand load can be estimated by reference to longshore transport experiments. A typical tracer advection velocity along shore is 0.5 cm/sec (Kraus et al., 1982), including movement by bed load and suspended load (a sand particle may experience both). Taking this value as an estimate, the time needed for green tracer to reach Station J, a distance of 15 m from the injection site, would be 50 min. Assuming that the fastest tracer grains move with double the above advection velocity would imply that small amounts of tracer may be expected to be found up-wave at sampling site J approximately 25 min after injection.

Figure 15.12 shows that a substantial amount of green tracer was found at Station J in the first sampling, 15 min after injection, implying an alongshore advection velocity of 160 cm/sec. Because the suspended load should move with approximately the velocity of the current (Figure 15.6), Figure 15.5 implies a short-term sand transport velocity of 50 cm/sec. Figure 15.6 may imply a 54 cm/sec offshore-directed flow in the region of Stations A and J. This result indicates that suspended load mode was probably dominant over bed load transport. The strong transport was localized as judged by the smaller amounts of red and orange tracer found on the grid and the close proximity of the injection sites.

DEPTH OF TRACER MIXING

The depth to which the tracer is found in a core can be determined according to the percentage of tracer included to that depth. It is assumed that a tracer is homogeneously distributed within each core segment. This model yields a percentage cut-off parameter that can be varied to examine depth of mixing as a function of concentration (Kraus, 1985). Two criteria were applied to define the depth of mixing. The first is that a minimum concentration must exist in at least one segment of a core. This minimum was set at 20 grains/100 g. A minimum-number criterion eliminates noise.

The second criterion applies the assumption that most tracer should be in the upper slices. If not, then accretion presumably took place. If less than 10% of the total concentration was not contained in an upper segment starting from the top, then that segment was not counted in the defining procedure. The zero mixing depth was then reset below the void segment. Both criteria serve to define depth of mixing in regions where substantial tracer was found. Finally, the average overall samplings at a given station were concentration weighted to express results as percentages.

With certain exceptions, most of the tracer was found to reside in the upper 4 cm of the bed. An interesting exception appeared at Station A (Figure 15.13). At the fourth sampling, a large amount of tracer seems to have been buried by 10 cm of sand. In the fifth sampling, the large discontinuity in concentration disappeared, and tracer was found to a depth of 12 cm. A similar departure was not evident at Stations

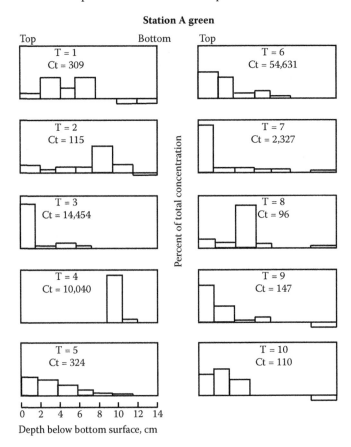

FIGURE 15.13   Example of depth of tracer mixing in bed (Ct = total tracer count in core).

B and C for the same samplings. Thus, the large accretion at Station A may have been a local process caused by, for example, a collapse in the rip channel bank.

CONCLUSIONS

This pilot experiment was designed (with great haste) to simultaneously measure waves, flow, and sand transport in a rip current. The main problem encountered was the physical limitations surrounding work in a strong offshore-directed current. With improvements in instrument reliability and decrease in their size, it appeared possible to make long-term point measurements of currents and waves in a large rip current. It was found useful to supplement the Eulerian current measurements with an overall view of current as indicated by the motions of dyes.

One significant finding was that sand sampling may be performed accurately and systematically in such a challenging environment. The key factors were the introduction of a new core sampler and sampling in a temporal arrangement. There is some doubt whether a single injection tracer experiment would reveal transport rates in a reasonable time span. Temporal sampling with continuous introduction of tracer may be more applicable (Duane and James, 1980).

The greatest transport occurred for the green tracer centered on the rip current axis, in contrast to transports by the feeder currents. Tracers closer to the feeder currents showed markedly less transport. The rapid spread of green tracer to all sampling stations indicates that suspended load was dominant in the rip. Suspended particles can be easily transported onshore by wave bores in addition to the rip current directed offshore.

Pulsations with periods of 45 to 60 min were found in the tracer motion and are attributed to long-period fluctuations in the incident waves and resulting current. Experimenters should be aware of this phenomenon when making short-term measurements of sand transport.

The depth of mixing varied from approximately 4 to 7cm, greater than mixing depths found in longshore sand transport experiments (Kraus et al., 1982). The greatest depth was found at the root of the rip where the feeder currents turn to flow offshore. Coupled with the reduced transport found at the feeder currents, the offshore transport by the rip current was primarily localized to the removal of sand along the rip channel.

ACKNOWLEDGMENTS

This chapter is the product of the Inlet Geomorphologic Evolution Work Unit of the Coastal Inlets Research Program, administered at the U.S. Army Engineer Research and Development Center, Coastal and Hydraulics Laboratory, for the Headquarters, U.S. Army Corps of Engineers (HQUSACE). It benefitted from reviews by Dr. Julie Dean Rosati, Ms. Tanya M. Beck, and Ms. Irene Watts. In writing this paper, I recalled mighty field assistance by colleagues in Japan: Dr. Susumu Kubota, Mr. Soichi Harikai, and Dr. Shintaro Hotta. Permission to publish this information was granted by HQUSACE.

REFERENCES

Arthur, R.S. 1962. A note on the dynamics of rip currents. *J. Geophys. Res.,* 67: 2777–2779.

Bowen, A.J. 1969. Rip currents 1. Theoretical investigations. *J. Geophys. Res.,* 74: 5467–5478.

Dalrymple, R.A. 1978. Rip currents and their causes. *Proc. 16th Intl. Conf. on Coastal Engineering,* ASCE, pp. 1414–1427.

Duane, D.B. and James, W.R. 1980. Littoral transport in the surf zone elucidated by an Eulerian sediment tracer experiment. *J. Sed. Pet.,* 50: 929–942.

Harris, T.F.W. 1967. Field and model studies of the nearshore circulation. PhD dissertation, University of Natal, Durban, South Africa.

Horikawa, K. and Hattori, M. 1987. The Nearshore Environment Research Center Project. *Proc. Coastal Sediments,* ASCE, pp. 756–771.

Hotta, S., Mizuguchi, M., and Isobe, M. 1982. A field study of waves in the nearshore zone. *Proc.18th Intl. Conf. on Coastal Engineering,* ASCE, pp. 38–57.

Komar, P.D. 1998. *Beach Processes and Sedimentation.* 2nd ed. Prentice-Hall, New York.

Kraus, N.C. 1985. Field experiments on vertical mixing of sand in the surf zone. *J. Sed. Pet.,* 55: 3–14.

Kraus, N.C. 1988. Measurement of sediment transport: sand tracer experiments. In *Nearshore Dynamics and Coastal Processes: Theory, Measurement, and Predictive Models,* Horikawa, K., Ed. University of Tokyo Press, Tokyo, pp. 433–439.

Kraus, N.C., Farinato, R.S., and Horikawa, K. 1981. Field experiments on longshore sand transport in the surf zone: time-dependent motion, on-offshore distribution and total transport rate. *Coastal Eng. Jpn.,* 24: 171–194.

Kraus, N.C., Isobe, M., Igarashi, H. et al. 1982. Field experiments on longshore sand transport in the surf zone. *Proc. 18th Intl. Conf. on Coastal Engineering,* ASCE, pp. 969–988.

MacMahan, J.H., Thornton, E.B., and Reniers, A.J.H.M. 2006. Rip current review. *Coastal Eng.,* 55: 191–208.

Sasaki, T.O. 1980. A heuristic model of the nearshore zone. *Proc. 2nd Symp. on Coastal Ocean Management,* ASCE, pp. 3197–3209.

Sasaki, T.O. and Horikawa, K. 1978. Observation of nearshore current and edge *waves. Proc. 16th Intl. Conf. on Coastal Eng.,* ASCE, pp. 791–809.

Shepard, F.P. 1936. Undertow: rip tide or rip current? *Science,* 84: 181–182.

Shepard, F.P., Emery, K.O., and La Fond, E.C. 1941. Rip currents: a process of geological importance. *J. Geol.,* 49: 337–369.

Shepard, F.P., and Inman, D.L. 1950. Nearshore circulation. *Proc. 1st Intl. Conf. on Coastal Engineering,* Council on Wave Research, pp. 50–59.

Sonu, C.J. 1972. Field observation of nearshore circulation and meandering currents. *J. Geophys. Res.,* 77: 3232–3247.

Yu, J. and Slinn, D.N. 2003. Effects of wave-current interaction on rip currents. *J. Geophys. Res.,* 108: 33-1–33-19.

16 Rip Currents
Terminology and Pro-Active Beach Safety

Stephen P. Leatherman

CONTENTS

INTRODUCTION

It is estimated that 100 people drown each year on United States beaches and probably thousands drown worldwide based on a review of the scientific literature (Klein et al., 2003; Hartmann, 2006; Short, 2007). Statistics from the U.S. Lifesaving Association show that approximately 80% of all lifeguard rescues at surf beaches are results of rip currents. Put into perspective, rip currents are responsible for more deaths than floods, hurricanes, or tornadoes on an annualized basis in the U.S. (Figure 16.1). Beachgoers presently have no method to directly detect and trace rips and other dangerous currents. This contributes to the high rates of mortality, near-drowning experiences, and needs for rescues at beaches around the world.

Red flags are used on many U.S. beaches to warn the public of marine dangers such as big waves, rip currents, sharks, and other hazards. Signs posted at beach entrances often contain idealized diagrams of rip currents, but rips take different forms and are thus unlikely to be recognized by the general public (Fletemeyer and Leatherman, 2010). Therefore, beachgoers often enter the water with little to no knowledge of these life-threatening currents (Figure 16.2).

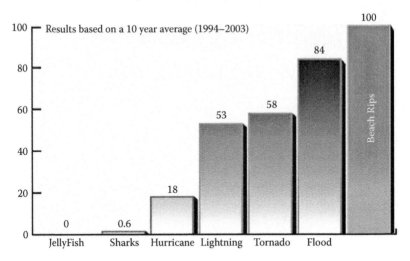

FIGURE 16.1   Rip currents, herein termed beach rips, cause more deaths on average than hurricanes, lightning, or tornados (not counting Hurricane Katrina in 2005).

FIGURE 16.2   Rip currents are often not apparent to beachgoers, and their identification can be problematic even for trained professionals. (*Source:* Rob Brander. With permission.)

Coastal researchers have used various water tracers by adding concentrated dye in powder or liquid form to water and then wading into the ocean and pouring the colored solution into the aquatic environment (Brander, 1999; Clarke et al., 2007). A more convenient way to deliver dye is in the form of a ball that can be thrown from shore. The dye ball dissolves to form a plume of colored water that delineates water currents. Although potentially very useful for coastal professionals (lifeguards, beach safety personnel, scientists, and engineers) in educational programs, product liability lawsuits represent potential problems for general public use.

Rip current terminology is presently not standardized in a societal context. *Undertow* and *riptide* terms are commonly used by the media and public to describe rip currents but are not accepted by coastal professionals. Even dictionary definitions are confusing and contrary to scientific definitions. A similar definitional problem existed for decades when tsunamis were commonly called tidal waves. In fact, tides have nothing at all to do with these earthquake-generated sea waves. The Great Tsunami of 2004 in the Indian Ocean—the deadliest in recorded history, causing catastrophic damage on several continents—was a game changer. The media and public lexicons changed instantly. Rip currents, by contrast, are localized problems, not world-shaking events. In fact, a rip drowning is often ignored by the media unless many individuals are involved. Conversely, the sighting of a single shark off a beach can cause panic and generate great photo opportunities and news headlines.

RIP TERMINOLOGY

The public has little understanding of rips, partially because so many names have been applied to these phenomena. The media usually refers to rip tides or riptides in describing drowning caused by offshore-flowing currents. *Rip tide* is a misnomer because tides play no role in generating rips—albeit such currents are often strongest near low tide. Along the New York Bight, beachgoers on Long Island and in New Jersey often call them *sea poosies*. *Run-out* is a common description, especially along the Florida Atlantic coast. Many bathers state that their greatest fear at surf beaches is undertow, which is not the same hazard as a rip current.

Every day some 6,000 waves break on an average beach. The water runs up the beach face as up-rush and usually stops near the berm, with some water soaking into the sand; the bulk of the water runs back down the beach face as backwash. Normally the return flow is fairly uniform along the beach so rip currents are not present. Big waves breaking directly on a beach generate a large up-rush and back-wash; this seaward-flowing water is also pulled strongly toward the next breaking wave. Waders feel as if they are being pulled under the water when the wave breaks over their heads—this is undertow in public parlance. While bathers can be tumbled around roughly, the return flow travels only a short distance—just to the next break-ing wave—and does not pull bathers offshore into deep water.

Coastal professionals who make presentations have long tried to eliminate the *undertow* term from the public lexicon with little to no success. In fact, anecdotal evidence indicates that such attempts have led beachgoers to discount the important information provided by coastal scientists. Instead of stating that *undertow* is a mythical term, which the public does not accept because of real-life experiences, perhaps it would be better to explain this current as strong beach backwash and sanction this specific use of the *undertow* term. Furthermore, this phenomenon must be differentiated from the much stronger and localized currents that flow seaward beyond the nearshore breakers—rips.

Coastal scientists (Shepard, 1936; Short, 1985; Leatherman, 2003) have consistently used *rip currents* to describe these rivers in the seas. The media and public prefer *riptides* perhaps because it sounds powerful, even though it is a misnomer. A riptide would technically be a tidal current associated with an inlet into a bay or lagoon; these currents, appropriately called *tidal jets*, can also be very strong and life threatening.

It is proposed that *beach rips* be considered as new and appropriate terminology for strong, offshore-moving currents generated on surf beaches and extending far into or through the surf zone. Interestingly, Short (2007) used the term to differentiate rip currents that occur in the surf zones of sandy beach and bar systems from topographically controlled rips and mega-rips. I propose that the *beach rips* term be applied to all rips that emanate from beaches. *Ocean rips* is not adequate because such phenomena exist offshore where strong wave and tidal currents interact at large sand shoals such as at Nantucket Sound, Massachusetts.

Beach rip terminology associates the action word *rips* with beaches, which will hopefully make the public take these killer currents much more seriously when bathing and swimming at surf beaches. George Orwell wrote that "If thought corrupts language, language can also corrupt thought." It is therefore important to get the language right; the standard dictionary definitions must be changed. No wonder the public is so confused by the varying definitions of water dynamics:

RANDOM HOUSE UNABRIDGED DICTIONARY (1987)

- Rip: a stretch of turbulent water at sea or in a river.
- Rip current: undertow.
- Riptide: a tide that opposes another or other tides, causing a violent disturbance in the sea.
- Undertow: the seaward, subsurface flow or draft of water from waves breaking on a beach. Syn. Riptide.

NEW OXFORD AMERICAN DICTIONARY (2005)

- Rip: stretch of fast-flowing and rough water in sea or river caused by meeting of currents.
- Rip current: an intermittent, strong surface current, flowing seaward from the shore.
- Rip tide: another term for a rip current.
- Undertow: current below the surface of the sea moving in the opposite direction to the surface current, especially away from the shore.

SCIENTIFIC DEFINITIONS

- Rip: same as rip current.
- Rip current: a strong, seaward-flowing current generated by waves breaking on a beach that moves offshore as a concentrated flow at all depths and extends into the surf (e.g., breaking wave) zone or beyond in some cases.
- Undertow: mythical current that does not exist in the surf zone.

PROPOSED DEFINITIONS

- Beach rip: a strong, seaward-flowing current generated by waves breaking on a beach that moves offshore as a concentrated flow at all depths and extends through the surf zone.
- Rip: same as rip current or beach rip.
- Rip current: same as beach rip.
- Rip tide: strong tidal flow through a constricted area, such as an inlet; often termed a tidal jet by coastal scientists.
- Undertow: strong return flow of water (e.g., backwash) from the breaking of large waves at surf beaches that does not extend beyond the next breaking wave.

It is interesting to note that the three most prevalent slogans used to raise public awareness of rips did not include the word *current*. A National Weather Service public relations campaign devised the "Break the grip of the rip" slogan. The Australian campaign as devised by Robert Brander at the University of New South Wales coined "Don't get sucked in by the rip." Finally, John Fletemeyer introduced "Don't get ripped" to the scientific community at the First International Rip Current Symposium held at Florida International University in March 2010.

A complicating factor in explaining the phenomenon of beach rips is that they cannot be characterized easily because of a diversity of shapes, sizes, and indicators of their presence in a surf zone (Fletemeyer and Leatherman, 2010). Some rips can be detected by lighter colored water because of bubbles and sediment (Figure 16.3). Others are delineated by darker water because of the presence of an underwater channel that is often controlled by bedrock (Figure 16.4) or reef breaks. The most common beach rips are fixed in place by low areas or holes in sand bars (Figure 16.5).

Flash rips are the most problematic as they are transient in nature and difficult to spot in a confused sea of breaking waves, often locally driven by strong onshore winds. Structural rips are permanently positioned by emplacement of shore-perpendicular engineering structures, especially groins, jetties, and even piers in some cases. These beach rips are caused by the seaward diversion of a longshore current (Figure 16.6). Finally, all these types of rips can pulse, making the current suddenly become much stronger because of the arrival of a large wave set (MacMahan et al., 2006). This pulsation is very hazardous for bathers and swimmers who sense little current; then the current suddenly becomes considerably stronger, sweeping unsuspecting bathers far offshore.

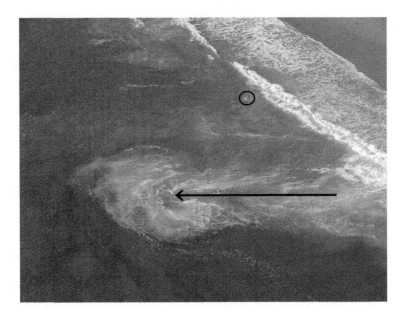

FIGURE 16.3   Rip at Zuma Beach, California is denoted by light brown water moving off-shore; note person in water. (*Source:* Los Angeles County Coastal Monitoring Network, 2002.)

FIGURE 16.4   (*See color insert.*) Rip currents are sometimes controlled by underwater rocks and reefs. The "Shell Beach Express" is exposed at spring low tide on a rare low-wave day at La Jolla, California. The Shell Beach rip occurs in the channel between two rocks as delineated by the elongated body of light-colored sand that is swept out when the current is flowing. Many bathers have been pulled offshore in this powerful rip current during big surf conditions. (*Source:* Stephen P. Leatherman.)

FIGURE 16.5 Family members enter the water at what appears to be the safest area because of a lack of wave action, but this is counterintuitive. In actuality, the gap in the bar is where a rip will occur under the right wave conditions. (*Source:* David Elder. With permission.)

FIGURE 16.6 Shore-perpendicular structures such as this groin at Cape Hatteras, North Carolina, often serve as pathways for rip currents. (*Source:* Stephen P. Leatherman.)

The purpose of suggesting this new terminology is to save lives through better communication of this dangerous but largely hidden hazard to beachgoers who bathe and swim at surf beaches. The first introduction of the *beach rip* term to the general public was through distribution in July 2010 of 60,000 copies of a four-page color brochure on rips that appeared as an insert in local newspapers in eastern Long Island (*Southampton Press* and *East Hampton Press*); it is also available at www. elicca.org. This new terminology was readily accepted by surf lifeguards and the generally well-educated and affluent public in the Hamptons.

The number of lives lost at surf beaches in the U.S. and Australia where the best drowning statistics are available shows little improvement in the past decade despite public relations campaigns by governmental and non-profit organizations (Fletemeyer and Brander, 2010). Clearly improvements in rip education are needed.

RIP DETECTION

Rip currents account for 80% of the U.S. Lifesaving Association rescues and are arguably the number one killer at oceanic beaches internationally; yet the public presently has no way to directly determine the presence of these life-threatening, offshore-flowing currents. Many rips are nearly invisible and are hard to recognize even by lifeguards. They are common in South Florida during strong onshore wind conditions. Beach safety personnel at Panama City Beach sometimes wade into the water to feel for rips; beachgoers generally have neither the swimming ability nor the training to undertake this kind of testing.

Warning signs and flags are used worldwide with varying success to alert bathers of dangerous water conditions, such as sharks, jellyfish, and rip currents. Expecting a bather to be able to identify rips using current signage and information is problematic, overly optimistic, and even unrealistic—the large number of drownings and rescues attest to this unresolved problem. "When professional lifeguards with many years of experience cannot always accurately identify the presence of a rip current, how can we expect the public to do the same?" (Fletemeyer and Leatherman, 2010). Current meters, GPS-tracked drogues, and microwave radiometers are used by coastal scientists and engineers to measure waves and water flows such as rip currents at surf beaches (Brander, 1999; MacMahan et al., 2006), but utilization of this instrumentation by the general public is not practical.

Dyes have been long been used by scientists and medical doctors as water movement indicators and blood flow markers in the human body, respectively. Coastal scientists use a number of dyes for water tracer studies, especially fluorescein, potassium permanganate, and rhodamine (Brander, 1999; Clarke et al., 2007), but fluorescein dye is the only one cited by the U.S. Environmental Protection Agency as safe in marine environments.

Fluorescein dye is available from a number of commercial manufacturers and vendors in the forms of tablets, liquids, powders, and solid cakes. These dyes are used only by professionals for specific uses and not by the general public. The tablet form is used by plumbers to trace leaks in pipes; the dye sinks to the bottom of a toilet bowl.

The tablet form is of no use as a surface water tracer. Liquid dye is concentrated and must be diluted by water. The powder dye is very light and easily blown by the

FIGURE 16.7 (*See color insert.*) Fluorescent green plumes from dissolving dye balls that move in a tidal current appear as white streaks in black and white. (*Source:* Stephen P. Leatherman.)

wind; it must be mixed with water before use in an aquatic environment. Diluted dye in liquid form is placed into a water column by researchers wading offshore into the current or may be thrown overboard from a boat. Powder and liquid forms of fluorescein dye can stain hands and clothes during pouring and mixing and cannot be readily injected into water currents tens of meters offshore; they are very inconvenient to use.

Oceanographic researchers have used solid dyes in waxy cake forms, placing them into floating casements to mark currents such as the Gulf Stream in the Atlantic Ocean. While the loss of a few such devices—usually made of Styrofoam for buoyancy—by scientists is not a significant problem, their widespread use by non-scientists would create substantial litter, much of it eventually washing up on recreational beaches.

The problem of introducing tracers into the surf was solved by manufacturing fluorescein dyes in solid forms so they can be thrown tens of meters offshore from beaches (patent pending by Leatherman, 2010). The dye balls must also float at the surface while dissolving to form colored plumes because only surface currents are readily visible by the naked eye (Figure 16.7).

Fluorescein dye is non-toxic, biodegradable, and NSF-approved as safe in drinking water (see Material Safety Data Sheet for fluorescein, disodium salt; www.hazard.com/msds). Dye balls are neutrally buoyant in water. They float at the surface but are not directly affected by the wind that would displace them relative to water currents. Fluorescein dye balls will hopefully prove useful by coastal scientists and lifeguards for public demonstrations and educational purposes. Testing by lifeguards at Miami Beach, Florida is presently underway.

DISCUSSION

Most warnings and signs depict classic rip currents that are perfectly formed with mushroom shapes (Figure 16.8). In actuality, rips cannot be classified easily because of the diversity of shapes and sizes. A wide range of indicators span the gamut from lighter to darker water, and rip currents cannot always be observed from beaches, especially when strong onshore winds are whipping up confused seas. Even lifeguards sometimes venture into the water to "feel" for rips that are difficult to see.

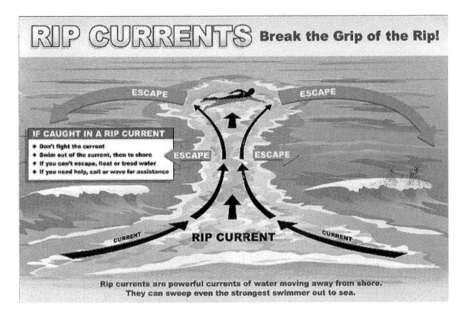

FIGURE 16.8 Typical warning sign for rip currents. (*Source:* National Weather Service and Sea Grant).

Safety signs are useful in depicting how a rip works and how to escape. The problem is that many beachgoers think that the mushroom-shaped diagram represents what they should see in the water. Therefore, current signage is not always helpful for spotting rips. Beachgoers need more information about waves and currents without having to become professional coastal scientists or oceanographers. YouTube videos that show rips in action are vitally important to demonstrate the presence and strength of these dangerous currents (www. Youtube.com + Rob Brander). A checklist has been designed to help beachgoers identify and hence hopefully avoid rips (Figure 16.9), but the public must be willing to spend a few minutes to make these observations before jumping into the water. Finally, water tracers are recommended to make both longshore and offshore-flowing rip currents visible. The problem with use of tracers by the general public is the possibility that some people will not realize that tracers show only currents at a particular location at the time of use. There are no safety guarantees as surf conditions change constantly and rip currents can be insidious due to their differing sizes, shapes, and strengths. This raises the issues of product liability and lawsuit potential.

Suits based on product liability represent a trend in the U.S., particularly in the areas of prescription drugs and medical care, and the courts tend to impose verdicts favoring plaintiffs because trial lawyers are adept at appealing to juries. This is especially likely when a child has drowned. Juries are unpredictable and can work more on emotion than scientific facts or practicalities. It can be difficult to determine how a case will be decided so insurance companies often settle out of court. The elements of product liability are: (1) duty—a product must perform as stated, (2) failure of a

- Listen to the surf forecast and watch videos of rip currents in action on YouTube before heading to the beach.
- Check for warning signs and flags; double red flags mean the beach is closed and a single red flag indicates no swimming allowed. Swim near a lifeguard if possible.
- Scan the water from the highest vantage point when you arrive at the beach and note the direction of wave approach.
  - If waves are coming straight onshore, there will be an increased danger of rip currents, but no current moving parallel to the shoreline (e.g., no longshore current). If caught in a rip, swim right or left (e.g., parallel to the shore) to escape this strong offshore-flowing current.
  - If waves approach from the north, the longshore current will be moving south. In this case, swim south with the longshore current and out of the rip.
  - If waves approach from the south, the longshore current will be moving north. Swim north out of the rip current.
  - The larger the breaking waves, the stronger the rip and longshore currents.
- Look at the line of breakers for a few minutes for any telltale signs of rip currents, which can vary by location:
  - Areas of less breaking wave activity (see Figure 16.2) where the rip is forcing its way seaward through the surf zone
  - Choppy water that extends beyond the breaker zone
  - Change in water color from the surrounding water—lighter color from bubbles and sediment (see Figure 16.3) or darker because of an underwater channel where the rip flows (see Figure 16.5)
  - Floating objects moving steadily seaward
- Rips frequently occur where shore-perpendicular structures such as groins, jetties, and piers direct the longshore current offshore (see Figure 16.6).
- Drainage pipes that empty storm water onto the beach and car ramps that channelize water to the beach result in rapidly-flowing water that cuts holes in sand bars; these areas are set up for rip currents.
- Rips tend to be stronger during times near low tide. Check for the high water mark on the beach as denoted by the wet/dry sand line, which is often the towel and beach chair line. At low tide, the high water mark is far up the beach face and the depth of water above the sand bar is minimized so people can often stand on it.
- Estimate the size of the waves, remembering that the power of a wave is proportional to the wave height squared; slightly larger waves are much more powerful (e.g., a 3-foot wave is 9 times more powerful than a 1-footer).
- Watch at least 10 waves in a row to see if the size is constant. On many beaches the waves build up and step down. Surfers sit on their boards waiting for the bigger waves to arrive so they will get the best ride. When these larger waves break onshore, they generate more powerful rip currents, which can pulse (e.g., double in strength and carry you farther offshore).
- Note the time in seconds between breaking waves;. The longer period of time between waves means that rip currents are more likely. Longer-period waves along the U.S. East and Gulf coasts are 8+ seconds, whereas Pacific coast waves normally have much greater spacing (e.g., 12+ second waves are particularly prone to powerful rips).

FIGURE 16.9   Rip safety checklist.

product to perform the duty, (3) negligence, and (4) injury to another party. These factors make the use of dye ball water tracers by the general public problematic. The issue of implied failure to perform the duty also extends to employing lifeguards on beaches.

Most beaches in Hawaii do not have lifeguards in spite of the big waves and powerful rip currents. The reasons are costs and liability arising from employing a guard if someone drowns or is injured (the state faces less liability for drownings on unguarded beaches). In many cases, Hawaii relies on beach safety literature placed in hotel rooms. This imposes the burden on visitors to become informed about water hazards. Even if visitors and tourists read and understand the written material and diagrams, their knowledge is usually not sufficient. Ballantyne et al. (2005) stated that "Awareness of rips and their potential dangers is worthless without an ability to recognize them, as only the latter can help to avert unsafe behavior." This makes the value of a water tracer apparent.

The Hawaii approach of dealing with rip hazards is replicated at many other U.S. beaches, such as Panama City Beach, Florida, but it hardly constitutes pro-active beach management as revealed by the large number of rip drownings. A stronger emphasis must be placed on pro-active beach safety and preventive practices, which should include more and better beach safety campaigns.

CONCLUSIONS

Beachgoers, especially visitors and tourists from non-coastal areas, have insufficient knowledge of rip currents. While waves at surf beaches are readily apparent, many tourists do not even know that currents exist, much less that they can be dangerous and even life threatening. The public education challenge is daunting, partly because our dictionary definitions are misleading and need revision to be scientifically accurate. Signs and flags are useful but not sufficient. Lifeguards are obvious solutions, but the cost is high and the U.S. shoreline is very long (e.g., Florida alone has 1,320 km of beaches). YouTube videos showing rip currents in action are vital because "seeing is believing," and are available instantly via the Internet. Miami Beach lifeguards are presently testing the dye ball water tracers; these public demonstrations will probably prove the best way for beachgoers to really understand near-shore ocean currents—both strong longshore currents and powerful, offshore-flowing rip currents.

ACKNOWLEDGMENTS

The Andrew W. Mellon Foundation and the Tom and Barbara Gale Foundation are gratefully acknowledged for supporting this rip research and educational initiative.

REFERENCES

Ballantyne, R., N. Carr, and K. Hughes. 2005. Between the flags: an assessment of domestic and international students' knowledge of beach safety in Australia. *Tourism Mgt.*, 26: 617–622.

Brander, R.W. 1999. Field observations on the morphodynamic evolution of low wave energy rip current systems. *Marine Geol.*, 157: 199–217.

Brander, R.W. and J.H. MacMahan. 2011. "Future Challenges for Rip Current Resarch and Outreach." In *Rip Currents: Beach Safety, Physical Oceanography, and Wave Modeling.* S.J. Leatherman and J. Fletemeyer, Eds. Boca Raton, FL: CRC Press, 1–29.

Clarke, L.B., D. Ackerman, and J. Largier. 2007. Dye dispersion in the surf zone: measurements and simple models. *Cont. Shelf Res.,* 27: 650–669.

Fletemeyer, J. and R. Brander. 2010. Presentations at First International Rip Current Symposium.

Fletemeyer, J. and S.P. Leatherman. 2010. Rip currents and public education. *J. Coastal Res.,* 26: 1–3.

Hartmann, D. 2006. Drowning and beach safety management along the Mediterranean beaches of Israel: a long-term perspective. *J. Coastal Res.,* 22: 1505–1514.

Klein, A.H.F., G.G. Santana, F.L. Diehl et al. 2003. Analysis of hazards associated with sea bathing: results of five years' work on oceanic beaches of Santa Catarina state, southern Brazil. *J. Coastal Res.,* 35: 107–116.

Leatherman, S.P. 2003. *Dr. Beach's Survival Guide: What You Need to Know about Sharks, Rip Currents and More Before Going in the Water.* Yale University Press, New Haven, CT.

MacMahan, J.H., E.B. Thornton, T.P. Stanton et al. 2006. Rip current review. *Coastal Eng.,* 53: 191–208.

Shepard, F. 1936. Undertow: rip tide or rip current? *Science,* 81: 181.

Short, A.D. 1985. Rip current type, spacing and persistence, Narrabeen Beach, Australia. *Marine Geol.,* 65: 47–71.

Short, A.D. 2007. Australian rip systems: friend or foe? *J. Coastal Res.,* 50: 7–11.

Index